PLAYING WAR

JOHN M. LILLARD

PLAYING

WAR

Wargaming and U.S. Navy Preparations for World War II

Potomac Books

AN IMPRINT OF THE UNIVERSITY OF NEBRASKA PRESS

Library of Congress Cataloging-in-Publication Data
Lillard, John M., 1958– author.
Playing war: wargaming and U.S. Navy preparations for
World War II / John M. Lillard.
pages cm
Includes bibliographical references and index.
ISBN 978-1-61234-773-8 (cloth: alk. paper)
ISBN 978-1-61234-825-4 (epub)
ISBN 978-1-61234-826-1 (mobi)
ISBN 978-1-61234-827-8 (pdf)
1. Naval War College (U.S.)—History—20th century. 2. War
games—United States—History—20th century. 3. United
States—Navy—Maneuvers—History—20th century. 4. Naval
strategy—Study and teaching—United States—History—20th
century. 5. Naval education—United States—History—20th
century. I. Title. II. Title: Wargaming and U.S. Navy
preparations for World War II.
V420.L55 2016
359.4'8097309041—dc23
2015029794

Set in Lyon Text by M. Scheer.

I dedicate this book to my family. First, thanks to my lovely wife, Barbara, who gave me the encouragement and indulgence necessary to make this quest a reality. Second, thanks to my sons, Jim and Sean. I went back to school for many reasons, but one of them was to show those two that it was possible for an old dog to learn a few new tricks. Last, I thank my dad, who I always felt was looking over my shoulder and enjoying what he was reading.

Contents

Illustrations

Tables

Acknowledgments

I particularly acknowledge the assistance of the Naval Historical Collection, especially head archivist Dr. Evelyn Cherpak, for providing me with the research materials and assistance necessary to complete this work and for never failing to come up with a resource when I needed it. Thanks also to Ed Miller, who many years ago guided and inspired me to take this project on and see it through to the end. I also acknowledge the support and assistance of my doctoral advisor, Professor Christopher Hamner, and my dissertation board members, Professors Martin Sherwin and Aric Thrall. Thanks also to a long list of friends, especially my study partners Becky Erbelding and Tracy Fisher, who always maintained a lively interest in this project. Finally, I recognize the assistance of the information technology section of Whitney, Bradley, and Brown Inc., who, more than they will ever know, facilitated the completion of this project.

Introduction

In October of 1960, Fleet Admiral Chester W. Nimitz delivered a speech to the staff and students of the Naval War College in Newport, Rhode Island, on how the "wargames" conducted at the War College during the 1920s and 1930s contributed to American tactical and strategic successes during the last two years of World War II in the Pacific theater. In his speech Nimitz, recalled by former director of naval history Earnest MacNeill Eller as the "principal architect of the American victory in the Pacific," asserted that "the war with Japan had been re-enacted in the game rooms [at the Naval War College] by so many people and in so many different ways that nothing that happened during the war was a surprise—absolutely nothing except the kamikaze tactics towards the end of the war."[1] This quote eventually became ubiquitous in histories, testimonials, and analyses of wargaming in general and the War College in particular. Comments by naval historians on the veracity of Nimitz's endorsement have run the full spectrum. Current Naval War College brochures print the quote without comment, Nimitz biographer E. B. Potter called the phrase "exaggerating a little . . . but basically tells the truth," Raymond Spruance biographer Thomas Buell characterized it as "dead wrong," and War College professor Douglas Smith wrote that the statement "could not be further from the truth."[2] From his position at the head of the Pacific Fleet during the war and as a War College graduate himself, Nimitz was certainly in a position to comment on the applicability of prewar training to actual combat. His confidence in the games' efficacy raises intriguing questions about the specific ways in which the games prepared and, in some respects, failed to prepare future naval leaders and their navy for war.

When Nimitz attended the War College from 1922 to 1923, the theories of Alfred Thayer Mahan still dominated U.S. Navy doctrine. Since 1890 Mahan emphasized that a great navy is one designed to fight an enemy in fleet engagements in order to win command of the sea, not one designed for commerce raiding or *guerre de course*.[3] This philosophy shaped U.S. Navy strategy and force structure well into the twentieth century. However, Nimitz and his classmates had never had the opportunity to put Mahan's theories to the test in actual combat. The last major American naval actions occurred against understrength Spanish fleets in 1898. American naval involvement in the First World War was limited to convoy support and antisubmarine patrols, and the minor naval operations of the interwar period, between 1919 and 1941, did not provide any occasions to prove or disprove Mahanian doctrines. But while their experience and education were grounded in the doctrines of the previous century, the student-officers of the interwar period like Nimitz, Ernest King, and Raymond Spruance led a very different navy to war in the 1940s. Their new navy possessed a long-range punch that centered on fast aircraft carrier task forces rather than battleship battle lines. This navy interdicted distant enemy lines of communication with high-endurance submarines. It repeatedly landed fully equipped ground units over enemy-held beaches, stayed at sea for long periods without returning to port, and carried its logistics support train forward with the battle fleet instead of sending combatant ships to homeports for refitting and repairs. It employed a wide range of new ship types and improved weapons, communications, and especially sensors such as radar and sonar. None of these capabilities were present in the U.S. Navy when the First World War ended, and none of the navy's leaders had much in the way of operational experience with them before Pearl Harbor. However, the navy introduced and experimented with all of them to varying extents during the interwar years.

These new developments were not just in doctrine and technology but in strategy as well. The navy's vision for a Pacific war evolved from a fast, decisive thrust across the ocean culminating in one climactic fleet engagement against the Japanese to a multiyear campaign with the final outcome dependent on a series of engagements, some decisive and others whose impacts would not be immediately apparent. American perception of the opponent expanded from a monolithic view of Japan as the sole adversary to a realization that the United States might face a multinational coalition of foes.[4] In a similar vein, the United States expanded its worldview to acknowledge both the need for joint operations with the army and air corps and the importance and contributions of allied nations who would be protecting their own interests in the Pacific.[5]

This catalog of progress raises a basic question: How did a navy whose last experience in major combat occurred in the previous century transform itself so profoundly? Nimitz's speech suggested that he credited the Naval War College's wargames with a major role in that transformation. Naval scholars generally agree that the Naval War College wargames were one of a number of techniques such as fleet exercises, formal war planning, and senior officer group deliberations that the navy used during the interwar period to manage its preparation for war. Comprehensive studies of the fleet exercises, formal war planning, and senior deliberations have informed the understanding of the navy's transition, but there is no analogous study of War College gaming during the period.

This book addresses that lacuna. It is a historically based inquiry into the behavior of a military organization, with specific attention paid to how that institution learns, evaluates and incorporates innovation, and transforms itself. The book describes how teachers and students came together to create and operate in an imaginary world made up of maps and models that represented what they thought the future might look like, to prepare themselves for that future. It investigates how the theory and practice of wargaming during the interwar years at the Naval War College affected players' perceptions of their real-world missions and prepared them for war by helping them to evolve beyond

Mahan to a doctrine that incorporated strategy, tactics, and technologies never used before in naval warfare. It joins an ongoing discussion of how military organizations evolve and transform over time by providing a historically grounded interpretation of the wargame phenomenon and its role in that transformation.

One of the persistent challenges facing military forces is maintaining proficiency during peacetime. Practicing the art and science of war is difficult enough at a tactical or small unit level. Exercising military decision making at the strategic level, where the considerations and conditions are considerably more abstract, is a much greater challenge. To help them teach strategic decision making, early military leaders reduced the complexities of strategy, logistics, and doctrine to simple games. The earliest examples of these were board exercises such as the King's Game of 1644 and military card games such as Le Jeu de la Guerre, played in France during the reign of Louis XV (1717–44).[6] From these modest beginnings, wargames grew in complexity as the decades progressed. By the second half of the nineteenth century, armed forces in Europe began to evolve away from their old militia models and toward the development of professional standing armies. These organizations featured permanent staffs to manage increasingly complex efforts such as long-range planning, procuring weapons, and supplying large armies in the field. This evolution increased the emphasis on formal military education, and thus the concept of wargaming evolved as well from a semirecreational board game to an officially sanctioned and encouraged training aid. The best known of these was the German Kriegspiel, first developed as a chesslike board game in 1780 and later superimposed over a map of the Franco-Belgian border. During the nineteenth century, Kriegspiel evolved to incorporate attributes of modern gaming such as scale distances instead of squares, sand tables instead of maps, umpires, pertinent scenarios, and elements of chance represented by dice rolls.[7]

European wargaming techniques found a receptive audience at the U.S. Naval War College. Founded in 1884 to "teach officers the science of their own profession," the college is the professional intellectual extension of the U.S. Navy.[8] Building on foundations laid by Mahan and War College founder Stephen Luce, the faculty introduced a systematic method of tactical analysis borrowed from the general staff of the German army. In 1887 the college took this emulation a step farther by incorporating a wargaming program adapted from Kriegspiel. The games in this program ranged from a single ship-versus-ship duel, through the fleet-versus-fleet tactical game, up to a strategic game that encompassed a complete theater of operations. The duel was of limited instructional value to a school that emphasized the study of fleet operations and quickly fell out of favor, but playing and studying the tactical and strategic games became central to the War College curriculum.[9] Staff instructors prepared the strategic game scenarios for the students, gave them overarching objectives based on the scenarios, and then divided them into groups representing both the United States and the opposition.[10] Both teams prepared plans to execute their respective strategies and issued orders for their forces. Team members playing the role of fleet staffs would plot ship movements on charts, while their opposition did the same in another room. This physical separation simulated the "fog of war," or the uncertainty of an adversary's capability and intent. As the opposing sides conducted scouting, game umpires who maintained their own "master plot" would progressively reveal more information to each side. Figure 1, an engraving by noted naval illustrator Rufus F. Zogbaum from an 1894 issue of *Harper's Weekly*, illustrates a chart maneuver of the period. Zogbaum, whose son later became a member of the War College staff, depicted a scene with a large-scale map of the Caribbean area hanging behind the students, who are working around a chart table mounted on sawhorses. The students are a mixture of navy and marine corps officers, with two observers from China in the background, watching the proceeding over the shoulders of their western contemporaries but not taking an active role themselves. The Chinese observers' position is a marked contrast to that of the officer in white in the foreground, who judging from his uniform and English-style rank insignia is probably one of the two Royal Swedish Navy officers who attended the course in 1894.[11] While foreign navy observers—

FIG. 1. An illustration by Rufus F. Zogbaum originally published in *Harper's Weekly* in 1894 showing a chart maneuver conducted at the Naval War College. Courtesy of the Naval War College Museum Collection.

including Captain Isoroku Yamamoto of the Imperial Japanese Navy in 1924—were a frequent sight at Newport during the interwar period, none of the classes during those years counted any foreign students in their rosters.

When opposing fleets closed within engagement range, the game venue transitioned from the chart to a larger-scale maneuver board with small models representing individual ships. Students maneuvered their ships by turns and exchanged fire with their opponents, while the game umpires evaluated hits and assessed damage. Sunken or damaged ships were removed from the board, and play continued until the instructors determined that the game objectives had been met. Reviews of student moves and instructor critiques followed the end of the game.

The War College games quickly captured the imagination of professionals and laymen alike. In 1895 navy secretary Hilary Herbert spent an entire visit observing game play, and came away from the visit visibly impressed.[12] After his own visit two years later, then assistant secretary of the navy Theodore Roosevelt wrote to War College president Captain Caspar Goodrich, "I look back with the greatest pleasure on my altogether too short of visit to the War College, and when I come on again I want to time my visit so as to see one of your big strategic war games."[13] Popular literature of the time also helped to fuel interest in naval strategy. Writers such as Homer Lea and Hector Bywater predicted in graphic detail a major war in the Pacific between the United States and Japan, and the maps they included in these fictional works found their way into many War College game scenarios.[14]

By August 1917 the War College had taken on the role of a laboratory for the development of naval war plans. In the years before there was a formal Navy Staff, a succession of navy planning boards in Washington (and on one occasion President Theodore Roosevelt himself) routinely submitted tactical, operational, and even technical problems to the War College for review or solution. Between 1890 and 1917, the War College student-officers prepared almost every war plan by themselves or in cooperation with the Office of Naval Intelligence.[15] The War College role as a war-planning body changed after the First World War with the establishment of the Joint Army and Navy Board, known generally as the "Joint Board," made up of the service chiefs and their senior advisors. The Joint Board organized a subordinate planning group and charged it with the development of joint service war plans. This Joint Planning Committee eventually subsumed the duties formerly executed by the individual service war colleges.[16] This restructuring of planning authority did not diminish wargaming activity at Newport. Instead, War College wargaming expanded and reached its zenith in the decades between the world wars, despite or perhaps because of the political pressures for disarmament and the economic strains of the Depression.

During the years between the world wars, navy leadership projected and prepared for the future through three basic venues and based strategic and policy decisions on those projections. The first venue encompassed the efforts of the Naval War Plans Division of the staff of the Chief of Naval Operations, referred to as OP-12. This group was responsible for the development of the naval aspects of the color-coded war plans developed by the Joint Planning Committee. The second venue was the annual fleet problems, which were live exercises of operating fleet units held between 1923 and 1940. The third venue was the wargaming program at the Naval War College. Throughout the interwar period, the work supporting these venues occasionally crossed (or did not cross) paths, either by official direction of navy leadership or more often by an osmosis-like process, as War College graduates moved on to fleet assignments and then back to tours in OP-12, bringing their previous experiences with them.

By the end of the interwar period, naval officers generally considered attendance at the War College to be a prerequisite for advancement to senior rank. Not every student who attended the War College became an admiral or a battleship commander, but a very high percentage of those chosen for senior leadership were in fact War College alumni.[17] By the start of World War II, 99 percent of all flag officers were graduates of the eleven-month course.[18] Of the forty names on navy secretary Frank Knox's 1942 list of the most capable admirals in the navy, thirty-six were those of War College graduates.[19] The same was true of every member of the naval contingent of the Joint Planning Committee and all the wartime fleet commanders. While these American leaders came from varied backgrounds and operational communities, they all shared two attributes. In 1941 none of them had any meaningful naval combat experience, but almost every one of them had practiced tactics and strategy in the War College's wargame environment. What effects did this shared experience in wargaming have on decisions made by these naval leaders once the real war started?

Historians of the interwar navy have examined four broad areas that reflect how the navy digested the impacts of the First World War and prepared itself for the Second World War. Those areas have aptly included the Naval War College in the list of institutions that contributed in some way to that preparation. However, in overlooking the evolution of wargaming during this period, previous scholars have ignored the transformational quality of the War College wargames and the central role they played in preparing the navy for war.

The first of the four areas of investigation, and the one of most direct applicability to this book, contains histories of the Naval War College itself. The second area contains studies of the army counterpart to the Naval War College, the Army War College in Washington DC. The third area examines the navy's annual fleet problems, the live-action counterpart to the War College wargames. The final area examines the Naval Headquarters, particularly the General Board and the War Plans Division of the Navy Staff. The studies that make up this historiography inform and provide a foundation for a detailed study of how the wargames prepared the U.S. Navy for the coming conflict.

Former director of naval history Ronald Spector made an important contribution to the history of the U.S. Navy in his 1977 book *Professors of War: The Naval War College and the Development of the Naval Profession*. Spector's history did not cover the interwar period but provided a detailed examination of the period between the founding of the college and the First World War, and a description of how wargaming came to be such an integral part of the curriculum. Spector's argument was that the War College both reflected and encouraged an increasing professionalization of the navy, a trend that was also seen in the other American services as they moved away from their pre–Civil War militia and citizen-soldier models. Spector's primary contribution to this study is his description of the genesis of wargaming at the War College and his documentation of its roots in the European Kriegspiel games. He used official correspondence to document how War College founders such as Stephen Luce and McCarty Little navigated a challenging course between the fleet, the General Board, and the Navy Department to establish a solid foundation for the War College as an intellectual center for the development of an educated, professional naval officer corps.

Spector ably treated the origins and developing mission of the War College, but there was an important oversight in his brief treatment and summary dismissal of the interwar period. In his epilogue, he looked forward from the end of his study scope to the 1970s and declared the War College program in the interwar period and subsequent years to have been "narrow, stereotyped, ritualized, [and] drained of relevance." His rationale for this severe judgment was the school's failure to aggressively experiment with new technologies such as submarines and aviation, and in their persistence in refighting fleet actions of the past such as Jutland.[20] Spector's review of interwar period wargames was brief and high-level (he covers the years between 1919 and the 1970s in seven pages), and he supported his assessments with a small sample of documents. By looking into the records and critiques from the wargames of this period in detail, *Playing War* provides a counterargument to Spector's assessment.

Naval War College professor Michael Vlahos's *The Blue Sword: The Naval War College and the American Mission, 1919–1941* is arguably the most in-depth study of the interwar period War College, but Vlahos's focus was not necessarily on the wargames and their preparatory and transformative role. *The Blue Sword* was a cultural history of how the War College contributed to the building and maintaining of a "warrior ethos" within the navy officer corps. Vlahos traced ethos (the collective identity) as a focal point to mission (the common vision of the future) to enemy (the inversion of the ethos), and interpreted the wargames as instruments of preparation for the mission. Like Spector, he largely dismissed the prewar games as highly scripted rituals whose primary purpose was to reinforce a warrior culture, dominated by a complex set of rules and managed by umpires he characterized as "Olympian overseers."[21] To trace the development of this warrior culture, Vlahos used a wide range of sources in the Naval Historical Collection, most notably the intelligence and technical archives, faculty and staff presentations, guest lectures, and student thesis and essay papers. He also illuminated his history with literature and quotations from much earlier periods, in an effort to display the roots of the maritime warrior culture as a separate and elite group in a larger, more common population. He did not use wargame records from the Naval War College archives.

An examination of wargame details and their progression during the interwar years was not Vlahos's intent, but if there was a gap in his cultural history, it is that he overlooked the wargames' pragmatic aspects. In the recurring play of the United States versus the United Kingdom situations, Vlahos saw the U.S. Navy constantly measuring itself against the "gold standard" of naval warrior ethos, the Royal Navy. In his search for deeper meaning, what he disregarded was the simple matter of the college staff's annual chore of developing challenging, semirealistic scenarios based on verifiable information. Vlahos quoted statements from senior officers like Holloway Frost regarding treaty inequities stemming from the Washington Naval Conference, and assigned much emphasis to an alleged inferiority complex of the U.S. Navy vis-à-vis the Royal Navy that War College game critiques and student writings do not reflect.[22] He similarly neglected the prevailing opinion that no naval professional of the time seriously thought the United States would ever square off against the United Kingdom.[23] As far as the games' role in this U.S-versus-U.K. fantasy showdown, he did not consider that students played *both* sides in the game, and played their assigned roles to the best of their abilities and not in accordance with how they thought their opponents would fight. Vlahos's examination of the games themselves occupied only twenty-five pages at the conclusion of his 161-page book.

Vlahos moderated his original dismissal of the wargames in "Wargaming: An Enforcer of Strategic Realism, 1919–1942", a *Naval War College Review* essay published six years after *The Blue Sword*. In this article he argued that the games did encourage an evolution in war plans and drove the development of an interwar period maritime strategy. He introduced the evolution of wargames through a three-phase progression that corresponded to explicit approaches, outcomes, and reactions to the gaming process.[24] Within that construct, he used a limited number of postgame critiques and special situation descriptions from one significant wargame in each phase to characterize the whole phase. The progression outlined in his essay and his method of using one significant game per phase as examples form the analytical framework for a much more detailed study in this book.

The most complete chronological history of the Naval War College is *Sailors and Scholars: The Centennial History of the U.S. Naval War College*, by John Hattendorf, Mitchell Simpson, and

John Wadleigh. First published in 1984, this survey of major events and developments throughout the history of the War College provides a useful backdrop to a more focused study. Hattendorf, a former professor of naval history and currently the director of the War College Museum, had complete access to the entire Naval Historical Collection, including a wide selection of photographs, and used these resources to good effect to document the progression of the college course of study and the place of wargaming within the curriculum. The review of interwar period wargames in this narrative is in some ways more comprehensive and detailed than Vlahos's treatment. The authors of *Sailors and Scholars* examined a somewhat larger sample of games, paid more attention to the role played by faculty and staff, and placed the game evolution in the context of what was going on with the college organization and in the greater navy at the time. If there is a deficiency in *Sailors and Scholars*, it is that this history was commemorative in nature and lacked critical analysis.

While histories of the Naval War College are central to the historiography of this book, histories of the advanced officer training in the army during the decades between the world wars provide a similar view of comparable institutions. The Naval and Army War Colleges had similar roles within their respective services and exchanged students and faculty throughout the interwar period. Several future World War II leaders such as Captains Thomas Hart and William Halsey attended the army counterpart after graduating from the navy school. This exchange was not simply a repeat of the same course of study in another location and from another perspective. The Army War College did not employ operational or tactical wargaming as part of its curriculum to the extent that the navy did, as the staff there considered their focus to be more on the strategic than on the tactical level.[25] Henry Gole's *The Road to Rainbow: Army Planning for Global War, 1934–1940* traced the Army War College role in preparing members of the War Plans Division of the Army Staff for war. Gole's argument was that the army began planning for a two-ocean war against Japan and Germany as early as 1934, and that a vital core of army strategic planners learned their trade at the Army War College. Gole used lectures, student papers, and class records from the U.S. Military History Institute together with War Plans Division records at the National Archives and Records Administration to show a much closer, official connection between the army staff planners and War College students than was the case in the navy. Gole intended his history to refute what he calls "mainstream interpretation" and "accepted historical wisdom" regarding the degree of army preparations for war, but he did not really specify any sources for that mainstream opinion. He did a thorough job of documenting the army's material readiness, or lack of it, during the interwar period but did not cite any recent historical scholarship that reflects an opinion that army strategic planning staffs were similarly unprepared.

Peter Schifferle's *America's School for War: Fort Leavenworth, Officer Education, and Victory in World War II* made the same claim for the U.S. Army's Command and General Staff College at Fort Leavenworth, Kansas, that Gole made for the Army War College. Schifferle's more recent history used an extensive collection of primary source material from the Leavenworth archives to place the Staff College in the forefront of institutions responsible for preparing the American military for the coming war. He described the Staff College as a school of division- and corps-level tactics that included a certain amount of wargaming in its curriculum, and the Army War College as a school of army-level strategy that did not.[26] One important aspect of Schifferle's book is how it contrasted the U.S. Army's comprehensive and multi-tiered officer training program with the U.S. Navy's single service college. Schifferle described how the Staff College made extensive use of practical exercises at Fort Leavenworth, but he made only two substantial references to wargaming and these described the army's program as narrower in scope and application than the Naval War College's.

The annual fleet problems figure prominently in studies of how the interwar U.S. Navy prepared for war, as they were live depictions of the same situations played out on the game board, and both were by nature somewhat inexact representations of reality shaped by evolving navy visions of what the future war would look like. Two recent histories of the fleet problems are *Testing American Sea Power: U.S. Navy Strategic Exercises, 1923–1940,* by Craig Felker (2007), and *To Train the Fleet for War: The U.S. Navy Fleet Problems, 1923–1940,* by Albert Nofi (2010). Each of these histories took its own approach in its examinations. Felker's argument was that the navy used the problems not to reinforce traditional Mahanian roles and missions but as operational experiences to modify their concepts of warfare.[27] He took a thematic approach based on technological and doctrinal innovations in his study, looking at the evolving use of aviation, submarine warfare, antisubmarine tactics, and amphibious warfare in the problems. Felker referenced the War College wargames only once in his text, simply echoing the assessments of Vlahos and Spector. Nofi, an analyst and wargaming specialist at the Center for Naval Analyses, took an alternative approach to Felker's focus on technological innovation by using a chronological approach to analyze the fleet problems in the context of the coming world war. He emphasized the "systemic interaction between the fleet problems and war gaming at the Naval War College," noting similarities in scenarios, record keeping, and umpiring guidelines.[28] He also pointed out that the same naval officer population participated in both fleet problems and War College wargames as both students and staff, and carried their experiences back and forth to each. Nofi gave particular attention to the relationship between umpiring rules developed at the War College and those used during the fleet problems, calling them "a subject in need of further study, as it involves the process by which lessons from wargaming helped shape tactics and doctrine."[29] While this book does not concentrate exclusively on umpiring rules, together with the rest of the wargame procedures they comprise a central part of this study of wargaming.

After wargames and fleet problems, the final leg of the war preparation triad is the efforts of the navy headquarters staff in Washington, especially OP-12 and the General Board. The most comprehensive history of the evolution of navy war plans is *War Plan ORANGE: The U.S. Strategy to Defeat Japan, 1897–1945,* by Edward Miller. Miller argues that contrary to contemporary briefs by army scholars who branded Plan ORANGE a defensive failure, the war plan was a valid, relevant, and successful guide to eventual victory in the Pacific.[30] Unlike Felker, and to a much greater degree than Nofi, Miller used the actual war plan documents (now part of the National Archives and the operational branch of the Naval Historical and Heritage Command) as well as records available at the Naval War College to relate War Plan ORANGE development with War College wargaming. He identified War College training as a "virtual prerequisite" for assignment to OP-12 and used communications from War College presidents regarding the merits of the different Pacific strategies to show an informal but close relationship between the college and the war planning staff.[31]

John T. Kuehn expanded on Miller's assessment of the General Board's role in interwar period preparation in his 2008 history *Agents of Innovation: The General Board and the Design of the Fleet That Defeated the Japanese Navy*. In a manner similar to Gole's approach with the Army War College, Kuehn intended to counter popular notions about the General Board as "hidebound, reactionary battleship admirals whose minds were closed to innovation" during the 1920s and 1930s.[32] Unlike Gole, Kuehn referenced the histories he intended to refute in his preface, and this thoroughness characterized the research supporting the rest of the book. He first discussed the mission and makeup of the General Board, especially its relationship with the staff of the Chief of Naval Operations (OpNav) and the War College. He documented how results of wargames were communicated to the board, and how the college in turn utilized the board's ideas in subsequent games. By emphasizing and documenting this open exchange of information and new concepts, Kuehn effectively refuted any notions that the General Board was resistant to change, and instead painted a picture of a receptive and influential leadership group that materially aided in the navy's preparations for the coming war.

In 2006 the father-son team of Thomas C. and Trent Hone contributed one more history to the ongoing dialogue of the interwar period navy, *Battle Line: The United States Navy, 1919–1939*.

As its title indicates, this book provided a much broader view of the navy as a whole, as opposed to an examination of one particular organization or aspect of it. Like Felker, the Hones organized their narrative thematically, and they focused on the testing and adoption of the same new technologies and doctrines that Felker reviewed in his study of fleet problems. However, *Battle Line* is not simply a repeat of *Testing American Sea Power*, as the Hones covered evolving tactics and the daily life of naval personnel to a level not seen in the other histories. The Hones gave more credit to the War College than did Felker, Spector, or Vlahos, calling it a center for innovative thought for both tactics and technology.[33] However, this opinion is tempered somewhat when one considers that the senior Hone is a former member of the War College faculty.

In total these histories by Miller, Nofi, Hone and Hone, Kuehn, and Schifferle used deep investigation of primary source material to rehabilitate the historical images of the engines of interwar period preparation in both the army and the navy. There remains only one major institution to receive this treatment—the Naval War College. Within his text John Kuehn provided a simple but eloquent Venn diagram of the four agencies that hosted the engines of preparation. Kuehn's diagram, replicated in figure 2, illustrates common membership (represented by overlap), communication, and coordination (lines) between the agencies.

If one expands on Kuehn's Venn diagram somewhat to include not only the different agencies and the fleet they supported but also the primary methods they used to prepare the navy for war, one sees that the War College wargames are the one mechanism not covered in detail by the existing scholarship (figure 3). Kuehn included the navy bureaus (Aeronautics, Ships, Ordnance, etc.) as a separate entity in his diagram, but these organizations managed design and procurement and did not have a significant role in operational planning and strategy development. Kuehn covers their activities in the area of preparations for the coming war in his history of the General Board.

While Vlahos and Spector covered the War College as an institution, what is missing from this historiography is a study

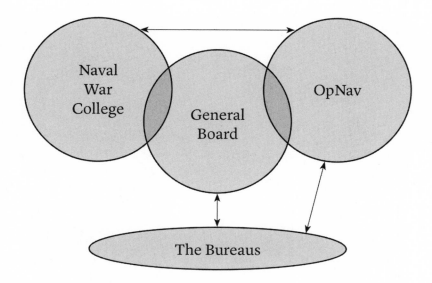

FIG. 2. A reproduction of John Kuehn's diagram showing the organizational relationships of war planning organizations during the interwar period. Kuehn, *Agents of Innovation*, 13.

that gives the Naval War College wargames the same in-depth review as war planning received from Miller, as fleet problems received from Felker and Nofi, and as General Board deliberations received from Kuehn. The purpose of this book is to focus on wargaming in detail and place it alongside war planning and fleet exercises as a method of preparation, a venue for experimentation, a palimpsest that recorded the progression of game development, and an instrument of transformation. It examines how the games evolved within the War College environment, how the participants understood the gaming experience, and how those experiences affected the outlooks, assumptions, and decisions of the commanders once the games were over and the real war had begun.

Tracing the evolution of a professional education institution and its primary instrument over a period of years brings with it the challenge of tracing different aspects of that institution that evolve in different ways and at different rates. In the case of Naval War College wargames, there are at least three aspects

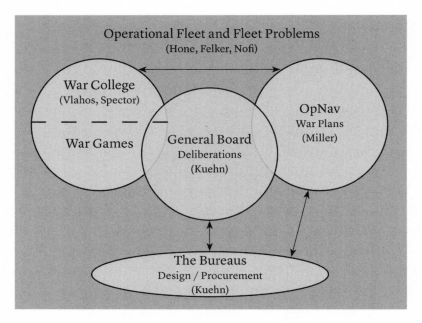

FIG. 3. An expanded view of John Kuehn's original diagram showing the organizational relationships of war planning organizations during the interwar period. Courtesy of the author.

to examine. The first aspect is that of the game *players*, meaning the staff and students who developed the game and gave it life. Game players hold primacy of place in this study, and this aligns with what wargaming expert Peter Perla describes as the War College's "underlying, almost subliminal, philosophy of never allowing the tools to dominate the [wargaming] process, of recognizing the ultimately central importance of the human being as both player and warrior."[34] A fresh batch of War College students arrived there every year and departed eleven months later. They entered with a certain level of knowledge and left having gained, along with their diplomas, a higher level of skill in decision making and a better appreciation of the navy outside their individual specialties. Their perspective of the school was more focused and short-term, and their evolution was more on a personal level. The college's perspective of students was necessarily similar. The student experience was not very different from class to class, and as a group, students left little behind in the way of lasting impacts. The case of the college staff is very different. Their time in the school was longer; one to three years per tour of duty, with some instructors returning for multiple tours. The faculty also left behind a corporate legacy of wargame scenarios, lesson plans, critiques, and the like for those that followed them. Individually, staff members evolved as their perspectives became deeper and broader, and the staff evolved collectively as the curriculum they managed matured.

The second aspect to examine is that of the game *process*, meaning the rules, equipment, facilities, and place in the curriculum. These elements persisted during the entire interwar period, evolving in detail but constant in purpose. Eventually, the overarching wargame process took on an identity of its own, and this process stayed remarkably constant when viewed in its entirety. While the game grew in scale, scope, and level of detail, a graduate of the class of 1919 who observed a wargame played by the class of 1940 would have recognized the game process immediately.

The third aspect examines the games themselves, meaning the scenarios, opposing sides, and objectives. These elements evolved with each play, with the results of each game absorbed by the students for the next game and processed by the staff for the next year. The scenarios changed to align with the world situation, the opposing sides adapted to accommodate real-world shipbuilding programs, and game objectives changed to align with new war plans.

Examining the history of the interwar War College in a purely chronological manner might obscure the differences in how each of these three entities—players, game process, and game results—evolved. Accordingly, this book employs a thematic structure to examine the players and the game process, and then a chronological structure to trace the evolution of the games themselves. The players are the subject of chapter 1, as understanding why the players were present at the War College leads logically to the role that the game process played as the centerpiece of the curriculum. Accordingly, the game process is the subject of chapter 2. Understanding that overall process in turn allows a deeper examination of the chronological progression of game outcomes,

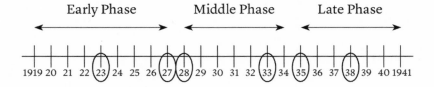

FIG. 4. A timeline of the interwar period, illustrating the three phases and six classes examined in this study. Courtesy of the author.

traced in chapters 3, 4, and 5. These chapters follow the phase convention used by Michael Vlahos in his 1986 essay "War Gaming: An Enforcer of Strategic Realism, 1919–1942" by dividing the interwar period at the War College into three phases.[35] Significant events of these years that directly influenced the context of wargames and the navy as a whole delineate these phases. In the early phase, from 1919 to 1927, the effects of disarmament, particularly the 1922 Washington Naval Conference, dominated naval strategy, tactics, and wargaming.[36] During the middle phase, from 1928 to 1934, increasing tensions overseas and the economic impacts of the Depression at home manifested themselves in gaming as well as the growing awareness of the role of new doctrines of naval aviation, submarines, and amphibious warfare. By the late phase, from 1935 to 1941, war was already under way in Europe, the U.S. Navy was in the midst of a major expansion, and U.S. war plans were in a tremendous state of flux.[37] Each chapter focuses on a series of wargames from two "significant" classes during each of these periods to assess game play, incorporation of new tactics and technology, and lessons learned. What makes these particular wargames significant or of value to this study is how they reflected or influenced evolving naval doctrine, influential members of the staff or student body that year, or both. Figure 4 illustrates how the three phases overlie the entire interwar period and where the classes of interest occur within those phases.

The primary sources that support this study are both official and unofficial. The official sources include the strategic and tactical problem descriptions developed by the War College staff, the written critiques and lessons-learned summaries delivered after each game ended, and other pertinent game-related documents maintained in the War College archives. This material includes class and staff rosters, textbooks and pamphlets that listed game procedures and rules, photographs of game facilities, and the charts and tables that documented game results. These official sources provide a continuum of results that track game evolution and the corresponding state of naval thought. The War College maintains extensive archives on the games in the Naval Historical Collection, a branch of the Naval War College Library. Two major record groups in this collection contain game documentation from the interwar period. Record Group 4, Publications, 1915–1977, contains scouting, screening, and operational problems; maneuver rules; quick decision problems; and game manuals. These documents are generally unvarnished views of game outcomes that the original writers did not plan to release outside the War College. One document type of particular interest is comments recorded during the critiques, which are not word-for-word transcripts but still capture some of the unfiltered give-and-take between professionals frankly discussing aspects of their profession. Record Group 35, Naval War Gaming, 1916–2003, contains maneuver board rules and manuals, fire effect tables and diagrams, syllabi and materials for wargaming courses, textbooks and histories of wargaming, correspondence, lectures, and classified game reports. Other official sources such as speeches or articles published in the Naval Institute *Proceedings* reflect either statements of purpose from the school or officer opinions on tactics and doctrine.

The unofficial primary sources include the writings of War College students and staff, particularly diaries or journals that some of them kept during their time at the school. Statements made in hindsight, like Nimitz's, are helpful, but observations made closer to the moment sometimes reflect a more candid opinion not necessarily modulated for public consumption. Personal papers can be double-edged swords to a historian, as they provide deep but narrow perspectives of events. The population of biography subjects and those whose papers merit archiving in a library can be self-selecting, which can likewise distort an attempt at a comprehensive review of a subject. To an extent, these attributes serve this book's purposes, as those who made the command decisions best answer an assessment of how the games prepared the navy for World War II. This investigation does attempt to go beyond the major commanders and examine the papers of staff members who observed and supported those commanders, as well as War College alumni who were too senior to participate in the Pacific offensive. Diaries are of particular value to this investigation, though it is of note that while many officers kept diaries, most did so only during their sea tours. Two major exceptions are Captain (later Admiral) Thomas Hart, graduate of the class of '23, and Commander (later Captain) Harry L. Pence '25, both of whom kept a daily journal through most all of their professional careers.

Other manuscripts referenced include personal papers at the Naval History and Heritage Command and the Library of Congress in Washington DC; oral histories maintained by the U.S. Naval Institute at Annapolis, Maryland, and Columbia University in New York; and a special collection of oral histories of members of the War College's permanent civilian staff. Francis J. McHugh recorded these individuals discussing the development and evolution of wargaming at the college from 1917 to 1965, with a focus on wargames conducted with Japan as the adversary.[38] These interviews provide a long-term and bottom-up perspective that is missing from those of the staff officers and students.

There are four points of clarification within this book regarding references to the War College alumni and the wargames themselves. First, to better relate the students to their time spent at the War College, graduates' names are appended with their class year when they are introduced, and their rank is given as it was during the period referenced. Accordingly, Fleet Admiral Chester Nimitz is referred to as Commander Nimitz '23 when his time as a student is discussed. Admiral Raymond Spruance is Commander Spruance '27 during his student year and Captain Spruance during his tour as head of the Tactics Department. Second, while the War College games are now and have been referred to in both official and unofficial literature as "games," the proper War College terminology for them is "maneuver," and that term was used most often, though not always, in War College official records. Third, there were multiple nomenclature and numbering conventions for maneuvers during the interwar period. The problems issued to each class were numbered serially with Roman numerals according to the class period in which they were used. Certain problems used from year to year were also given permanent Arabic identification numbers followed by modification numbers in case the same problem was used in an altered form.[39] The references within this book adhere to the wargame designations that appear in archived class schedules, maneuver histories, and critiques. Fourth, the actual maneuver schedule throughout a student year varied from the written schedule as necessary. Instructors continued a given maneuver until they believed that the class had achieved their learning objectives, and then the game was ruled complete. The maneuver might be started again depending on other class activities, and it might also be played at different times by different groups of students. Dates on the maneuver records can indicate the date that the document was prepared, or the "game date," as opposed to when the maneuver was conducted, and while these might align somewhat, the upshot is that compiling a comprehensive chronology of maneuvers presents a challenge. The most complete list of interwar period maneuvers was included as an appendix in Vlahos's *The Blue Sword*. That list forms the basis of the catalog included in appendix B of this book.

1

The Players

When Lieutenant Commander Chester Nimitz reported for duty at the War College in 1922, like all of his classmates he commenced what would be an atypical year in the career of a professional naval officer. Rather than leading subordinates in the performance of duties that he understood thoroughly, Nimitz would be cooperating with his peers to address problems and expand their knowledge in areas where they had little prior experience, using a method that most of them had never seen before for simulating naval combat. After graduation, they would leave the War College and return to familiar fleet or staff positions but with their personal and collective perspectives on strategy and tactics transformed to varying extents.

To better comprehend the transformative effect the wargames had on the navy, we need to understand the people who lived through that transformation. Accordingly, this study of wargaming in the interwar period begins with an examination of the players who participated in it—who they were, where they came from, how they came to be at the game, what their surroundings were like, and which preconceptions they brought to it. In this case, "players" refers to the staff who ran the War College, the faculty who orchestrated the games, and the students who acted out their assigned roles on the opposing sides. Previous historians like Spector have used vocabulary like "narrow, stereotyped, ritualized, and drained of relevance" to describe the interwar period wargames. If this were in fact the case, it would not be surprising to find primary source evidence to support such a claim. There was no one closer to the games than the staff, faculty, and students. If the interwar period wargames were irrelevant rituals, one would expect there to be some evidence among the postgame critiques and memoirs that reflected a War College environment that was not conducive to experimentation or learning. Such evidence might include rigid mission statements, outdated course material that did not incorporate new information, inadequate or ineffective staff, one-way communication between faculty and students, highly structured student routines, or student memoirs that imply that the wargaming experience was a waste of time or that the environment left little room for independent thought.

Because of their different roles and their longevity at the college, staff, faculty, and students had very different perspectives of the games, their purpose, their effectiveness, and their downstream effects on naval preparation for war. This chapter examines these different perspectives. The War College staff set the objectives and administered the curriculum for each group of students; accordingly, this chapter first examines the college's mission and vision, how the faculty went about accomplishing it, and what sort of environment they created while doing so. It then looks at the students: their backgrounds; how they came to the college; how they lived, worked, and learned within the War College environment; and how they perceived their experience both during their time as students and with the benefit of hindsight.

The interwar period at the Naval War College began with the return of Admiral William S. Sims to the position of school president in 1919. Sims had to cut short his first tour as president in 1917 when he was ordered to England, first as a senior naval observer, then as the commander in chief of all American naval forces in Europe, with the rank of vice admiral. Sims distinguished himself as a wartime leader, and the fact that he voluntarily accepted a reduction in rank to return to his peacetime assignment provided the college with a considerable amount of publicity and prestige. He became the highly recognizable face of the college for the first years of the interwar period.

Sims maintained a consistent vision of the War College as an institution that would carry on the school's original Mahanian objective, which Mahan himself spelled out at the opening of the War College's fourth session. Mahan declared the War College to be an institution that would "promote, not the creation of naval material, but the knowledge of how to use that material to the best advantage in the conduct of war."[1] However, despite Mahan's imprimatur and Sims's formidable presence, the "greater" navy did not fully accept the War College as an essential part of a naval officer's career development in the beginning of the interwar period. In a holdover from the prewar era, senior naval leadership of the time valued the "practical man" more than the "theoretical" and believed that the place for a naval officer's formal education was aboard ships and not in the classroom.[2] In the greater navy, this philosophy flowed down from senior leaders to junior officers. Accordingly, Sims's challenge was two-dimensional: to bolster the school's relevance in the eyes of the greater navy while at the same time establishing an environment inside the school where the proper combination of theoretical and practical learning could flourish. Sims's ability and authority was such that he was eventually able to realize his vision and have it maintained unchanged for twenty-two years.

Sims expanded on Mahan's interpretation of the War College mission in a speech to the officers of the U.S. Naval Academy on 11 November 1912. He stated, "The primary objective of the Naval War College is to study the principles of warfare . . . to develop the practical application of these principles to war on the sea under modern conditions, and then to train our minds to the highest degree of precision and rapidity in the correct application of these principles." Later in the speech, Sims elaborated on what he called "wholly essential" qualities necessary for a naval leader to apply these principles properly. He said, "These qualities . . . comprise the ability to recognize . . . promptly, the military significance of each strategical and tactical situation; the ability to withstand surprise without impairment or suspension of judgment; rapidity of decision and promptness of action; and inflexible determination in carrying out the plan of operations."[3]

The recognition that Sims emphasized he repeatedly referred to in his speech as the "estimate of the situation." This phrase refers to the act of processing available information and determining a course of action. In other words, Sims called the mission of the War College to teach officers *how* to think, not *what* to think. The estimate of the situation became the foundation of a deductive system of studying and solving war problems, and that system became the focus of the War College curriculum.[4] The college formally documented guidelines for developing estimates in a pamphlet in 1910, and later expanded the pamphlet into the official War College publication *The Estimate of the Situation: Plans and Orders* (later revised into *Sound Military Decision*).[5] *The Estimate* evolved into a guidebook by which U.S. naval officers learned to dissect and diagnose naval problems. The War College staff never intended *The Estimate* to be a rulebook or set of procedures, however, and the official college philosophy toward such an interpretation was explicit. As was stated in the booklet *The Mission and Organization of the Naval War College, 1936-1937*, "Human action cannot be governed, nor can war successfully be waged, solely by precedent or by adherence to rule. The College offers no rules for the application of fundamentals and sedulously advises avoidance of such rules. Development of sound professional judgment, through unremitting individual study and observation, is the only path to the successful application of fundamentals. Assistance to the individual in this development is the offering of the College."[6] This philosophy of "assistance but not rules" is crucial to understanding the War College course of study construct and how the wargames fit into it as laboratories for practice and experimentation. While the curriculum matured and instructors came and went, the Estimate of the Situation—both the publication and the action—remained foundational. Naval War College students went to classes to learn about it, attended lectures to put it in the proper context, and practiced it repeatedly during the wargames that formed the major part of their curriculum.

Sims viewed the games from the perspective of a practical man. To him, they were simply a form of practice for developing estimates, in the same way that warship commanders of the time used "dotter" and subcaliber exercises to train their gun crews.[7] Sims understood that opportunities to conduct full-scale fleet maneuvers would be few and far between but that repeatable wargames could provide inexpensive and accessible opportunities for student-officers to learn through trial and error that which they could not hope to experience at sea. Nevertheless, while Sims stressed this practical aspect of wargaming, he also could see that the games could "serve to develop new applications of the principles of warfare as applied to modern naval conditions."[8] His qualification left the door open for the War College to move beyond simple training and into the realm of strategy, tactics, and technology development.

Learning how to properly assess situations, develop orders, and experiment with new applications was central, but Sims was adamant that the War College not have any direct role in the development of official war plans. In a letter to the secretary of the navy, he stated, "It should be well understood by the service that the college is in no sense a planmaking body, nor has it any administrative or executive functions."[9] This policy, codified in the *Mission and Organization of the Naval War College* pamphlet, echoed Sims's vision of the War College as an institution that provided "an uninterrupted opportunity, free from administrative demands, for concentration [on study of the exercise of command]."[10] This was one guiding principle that the college maintained (with one exception during the presidency of William Pratt) throughout the interwar period.[11] The last president during peacetime, Edward Kalbfus, confirmed the durability of this philosophy in a statement that Sims would have endorsed, describing the college as "existing for the mental advancement of the individual student officer, not as a reference point for profound opinions, nor as a test plant for war plans, nor as a proving ground for suggested new types."[12] This position relative to the activities of the "greater Navy" led to something of a dichotomy. While there was no *direct* connection between the college and OP-12 (the War Plans Division of the Chief of Naval Operations' staff), there was an indirect connection through a sort of cyclic osmosis. Students arrived at the War College from the operational navy having made deployments overseas and having participated in the annual fleet problems, which were field tests for the various war plans. These students rotated back to the fleet, carrying their War College experiences with them. Some students became instructors who wrote the wargame scenarios set in real-world geographic areas and could not help but be influenced by their own experiences in the fleet and by current thought among members of the navy staff.[13] In fact, attendance at the War College was a prerequisite for assignment to OP-12.[14]

The historiography of the War College tended to emphasize the role and influence of War College presidents. While the president was the most visible face associated with the college, wargame records reflect that the rear admirals who occupied the president's position actually had much less to do with the wargames than did the captains who chaired the Operations, Strategy, and Tactics Departments. This latter group of officers directed the execution of the curriculum, designed the wargame scenarios, approved the class assignments, officiated at the wargames, and led the postgame critiques. Accordingly, it was they, and not the War College presidents, who were most responsible for the degree to which wargaming impacted preparation for real war.

Admiral Sims's status, reputation, and the loyalty he inspired in officers who had served with him previously attracted high-quality instructors to the school.[15] Staff rosters from the first years of the interwar period include officers like Joseph Taussig and Reginald Belknap, who had distinguished themselves in the U.S. Navy's limited combat operations during the late war.[16] Sims was also able to persuade notable figures from government, business, and academic circles such as James Quayle Dealey, chair of the political science department at Brown University, to contribute regularly as guest lecturers. As the interwar period progressed and Sims's handpicked staff rotated back to fleet assignments, new staff instructors were drawn from the best and brightest of War College graduates. At the end of their term, former students such as Harris Laning '22, Carl Moore '35, and Bernhard Bieri '36 simply changed sides of the classroom to become instructors. Some like Thomas Withers '24 and Raymond Spruance '27 came back for multiple tours. These instructors would have major influence on how the wargames evolved over the interwar period.

While the drawing of instructors from a small pool of distinguished graduates might seem like a closed-loop system that would stifle original thinking, a number of interwar instructors were notable for how they modernized the curriculum and updated the maneuvers. One of the best examples of this was Harris Laning, who served two tours on the staff, first as head of the Tactics Department from 1923 to 1924 and next as president from 1931 to 1934. Today Laning is an obscure figure in naval history, as he was assigned to the Bureau of Navigation during World War I and was too senior to participate actively in World War II, but his position is much more prominent in the more esoteric histories of naval wargaming. Both John Hattendorf and Albert Nofi give him a large share of the credit for injecting discipline, rigor, and relevance to wargames of the interwar period.[17] Laning's flag lieutenant, James Holloway, recalled him as an innovator who advocated for new ship formations, building air-capable ships, and placing the fleet commander out of the battle line, either on board a specially equipped flagship or ashore.[18] One of Laning's first significant contributions to War College wargaming was his senior thesis, "The Naval Battle," which became a textbook for subsequent classes. He also exhibited more flexibility than the stereotypical interwar naval professional might be given credit for. After the class of 1923 (Laning's first class as an instructor) finished two rough performances in their major tactical maneuvers, his students requested an outline that might provide more specific guidance than *The Estimate of the Situation*. Laning responded with a fifteen-page how-to document titled "Hints on the Solution of Tactical Problems."[19] The procedures he outlined in this paper found their way into the subsequent update of *The Estimate of the Situation*.

Two of Laning's students from the class of 1924 gained reputations as advocates for the navy's nascent aviation and submarine forces. Captain Joseph Reeves served on the college staff as head of the Tactics Department in 1925 before becoming one of the first senior naval officers to be designated as an aviator. Reeves's biographer, Thomas Wildenberg, credits him with developing the carrier task force concept and putting it into practice during fleet problems.[20] Reeves's classmate Captain Thomas Withers was an early advocate for technology advances in submarine habitability and endurance, and promoted the long-range interdiction tactics that the submarine force would eventually employ with such success during World War II in the Pacific theater.[21] Withers served on the tactics and operations staffs from 1924 to 1926 and later returned in 1928 to manage the junior course.

William Veazie Pratt was not a War College graduate, but he did serve a tour as an instructor before his appointment as president in 1925. Pratt was both an intellectual and something of an anomaly among his fellow flag officers because of his views in favor of disarmament and his work in support of the Washington and later London Naval Conferences.[22] While he was not completely convinced of the practicality of submarines in a naval campaign, Pratt was very clear-eyed regarding the possibilities of naval aviation, and his statements on the subject are a solid counter to the popular vision of interwar leaders of the navy as hidebound traditionalists.[23] When he arrived at Newport, he immediately emphasized the neglected field of logistics at the college and aligned his instructor staff to mirror the organization of a fleet staff.[24] He also attempted to restore the college to its original role in the war planning process, but by the mid-twenties, the separation between the college and OP-12 was too deeply institutionalized. Both Ronald Spector and Michael Vlahos emphasized Pratt's highly visible and ultimately short-lived attempts to realign and reinvent the War College as examples of interwar navy resistance to change, but these perspectives overlook a steadier and long-term evolution reflected in wargame records. Logistics planning remained in the curriculum after the functions of the Logistics Department were absorbed into those of the Operations Department, and the flow of War College graduates to OP-12 continued.

Raymond Spruance '27 is probably the best known of the interwar instructors. He maintained a long relationship with the col-

The Players

lege, both as a student and during three tours on the staff, and remained an influential advocate of the school's mission and philosophy throughout his career.[25] According to postwar recollections of his faculty colleagues and students, Spruance's primary contribution was to set an atmosphere in which both students and instructors were free to express their opinions, to innovate, and to experiment.[26] Commander Carl Moore, Spruance's assistant for tactics in 1937 and later his wartime chief of staff, recalled him as "liberal . . . allowing me to do almost anything I pleased."[27] Spruance's subordinate for strategy in the Operations Department was Richmond K. Turner '36, the future commander of amphibious forces in the Pacific War. Many of Turner's fellow officers recalled him as a hard-nosed and difficult man, but they also remembered him as an innovator and a relentless advocate for a greater role for aviation in the navy. As he did with Carl Moore, Spruance gave Turner free reign to update wargame scenarios, improve staff presentations, and emphasize the role of aviation and amphibious operations in the games.[28]

The staff and faculty of the War College developed and managed a curriculum that defined the breadth and depth of the educational experience there. While wargames were the centerpiece of the curriculum, they were not the only component. During the first months of their course of study, the students heard a battery of classroom lectures led by the War College faculty. The students augmented this theoretical foundation with independent study before they moved on to their practical exercises. Students attended classes on command, strategy, tactics, and international law. Instructors introduced the estimate of the situation—both the theory and the outline itself—during the command class. The strategy and tactics classroom sessions eventually gave way to maneuvers as the year progressed. Reading lists for these classes were extensive and mixed classic theoretical works on strategy and tactics by Thucydides, Mahan, Clausewitz, Corbett, and Darrieus with more contemporary accounts from both sides of Great War naval actions by Jellicoe and Scheer.[29] Students plumbed these readings for foundational strategic and philosophical insights, and not fleet tactics. For example, a common theme throughout these readings was the importance of maintaining the offensive. War College faculty used these readings to promote a philosophy that emphasized taking and keeping the initiative in battle, and references to this philosophy were prominent in wargame critiques.[30]

Students also heard guest lectures every Friday during the first six months of the course, and then less frequently once the wargaming phase received more emphasis in the second half.[31] Most of the naval lecture subjects included both broad topics such as aviation and submarines and more specific subjects such as late developments in armor. The lecturers themselves were not student peers or even staff instructors but were, instead, authorities in their respective fields.[32] Viewed from today's perspective, these guest lectures reflect recent employment in the First World War or contemporary doctrine and technology as opposed to pending development. The impact and most probably the intent of these lectures then was not to disseminate news of *pending* technical or tactical developments but to familiarize students with the *current* characteristics and *existing* capabilities of other naval branches.

War College students also heard lectures on political science and sociology subjects with such titles as "The World Situation as It Affects the United States" and the "Relationship of Geography to the Character of People of the Far East."[33] Like the professional course material, the content of these lectures themselves was not innovative or transformative, as they largely reflected the prevailing Anglo-Saxon perspectives of politics and the rest of the world. If there is innovation here, it is not in the material itself but that such subjects were presented at all, in an effort to broaden the student-officers' perspective and enable them to place their naval tactics and strategy into some sort of geopolitical context. There was no other official venue in the entire navy for naval officers to receive such information formally.

Across the classroom from the staff and faculty sat the War College students. Considering that they all came from the same population of naval professionals and were only in place for eleven months on their way to other assignments, this group of officers might seem like unlikely prospects for transformative thinking. Yet they experimented with new tactics and technology, attempted different strategies, and engaged in spirited debates over wargame results with their instructors. Those who returned to the college as faculty members incorporated their experiences as students into an evolving curriculum. How did they absorb and incorporate so much new information without the benefit of learning it at sea in actual operations?

Part of the answer lies in their unique culture. The naval officer community during the interwar period was an extremely homogenous group; close in economic circumstances, education, and political outlook.[34] The student-officers at the War College— uniformly white men—came from all states of the union and from a variety of backgrounds to join a rigidly defined military hierarchy that emphasized deference to seniority. Their shared experiences after they joined the service defined them as a separate culture that was distinct not only from the general population but also from the other military services. The naval officers that comprised the majority of the War College student body all received their commissions from the same school—the U.S. Naval Academy at Annapolis, Maryland.[35] Once the postwar drawdown was complete, non-Academy graduates commissioned into the Reserves went back to their civilian lives, and those granted temporary commissions reverted to their enlisted rank. What remained was a naval officer corps made up almost exclusively of Annapolis graduates.[36] By contrast, the U.S. Army officer corps in the interwar period included a higher percentage of former enlisted men and graduates of colleges other than the Military Academy at West Point, New York.[37]

After graduation from Annapolis, naval officers had much broader international exposure than did their counterparts in the Army. A typical naval officer had multiple opportunities to visit other countries and interact with foreign cultures as a routine part of his sea duty assignments. Naval War College student and staff memoirs are replete with discussions of port calls in South America, Asia, Europe, and the Mediterranean. Junior officers who accompanied Theodore Roosevelt's "Great White

Fleet" on its 1908 round-the-world cruise would become prominent World War II leaders.[38] By comparison, an army officer's opportunities to serve overseas during the interwar period were typically limited to multiyear assignments pacifying insurgents in the Philippines or guarding the foreign legations in China.

Regardless of their individual specialties, collectively the interwar period naval officers were sailors in the traditional sense. Their professional world was that of ships at sea, and the small number of civil engineering, supply, and medical corps officers who attended the War College were no exception. There were few submariners and even fewer aviators in the student ranks. Officers who qualified in submarines did not necessarily spend their entire careers in that community. Capital ship command was the path to flag rank, so it was not uncommon for battleships or heavy cruisers to be commanded by officers who wore the twin dolphin insignia of a submariner (e.g., Nimitz and Thomas Hart '23).[39] Between 1930 and 1935, only three classes included naval aviators and there was only one naval aviator on the War College staff.[40] Senior officers such as Joseph Reeves '24 and William Moffett, who passed the aviation observer course, took command of the new aircraft carriers, but they were neither career aviators nor even qualified pilots.[41] As late as 1936 Commander Bernhard Bieri admitted to coming to the War College with little to no knowledge of aircraft carrier or submarine operations.[42] However, while War College students might have lacked a broad understanding of emerging naval branches, they were not unschooled. Their sea duty assignments required deep technical expertise across a range of disciplines. Nimitz was the navy's acknowledged authority on diesel engines.[43] Husband Kimmel '26 was an expert in the mathematics and physics of gunnery.[44] Spruance spent a year studying electrical engineering at the General Electric Company.[45] Benjamin Dutton Jr. '29 was the author of a series of maritime navigation texts that still bear his name today.

Naval officer training was not limited to technology. Student writings, both for class assignments and in their personal journals, reflect an awareness of and appreciation for the history of their profession. War College instructors fed this interest by frequently emphasizing points in game critiques not only with references to the comparatively recent battles of Coronel, Falklands, and Jutland but also with historical allusions to or case studies of Lake Champlain, Trafalgar, Gettysburg, and other battles fought long before the students were born.[46]

Common experiences gave the student-officers a foundation that made them used to collaborative learning. Their deep training in engineering and mathematics showed them the power of quantitative data and scientifically derived results. This appreciation for objectivity made the students receptive to new concepts, even transformation, as long as they saw the supporting evidence. As student-officers participated in the wargames, reviewed the results, and developed lessons learned as a group, the transparency of the wargame process and the preponderance of wargame results provided that quantitative evidence.

The Bureau of Navigation, the agency responsible for duty assignments and other personnel-related issues, selected officers to attend one of the one-year courses at the War College. The bureau based its selections on a number of criteria that were not always consistent. In the constant jockeying for senior command and flag rank, naval officers variously considered a War College tour to be a necessity, a place to wait while another billet opened up, or a dead end. Most officers actively sought War College orders, but others came to Newport reluctantly, already holding a low opinion of the school and its graduates.[47] Most students had to relocate from distant places in one of those frequent moves that still characterize military life.

Once students arrived at the War College, their shared experiences continued in an environment designed to remove them, as much as possible, from the day-to-day responsibilities normally associated with a navy assignment. Academics were the student-officers' sole professional concern during their time at the college. They had no watches to stand, no sailors to supervise, no inspections, and none of the other duties associated with shipboard life.[48] Civilian clothes were the uniform of the day, to include rubber-heeled shoes to keep noise down in the hallways and to prevent marks on the maneuver boards.[49] Not wearing uniforms meant dispensing—temporarily—with visual symbols of rank, service, and seniority, which in turn encouraged the students to interact with each other on a more level basis and conduct a freer exchange of ideas. This exchange was evident when junior and senior classes combined in some of the larger games. Transcripts of critiques from these games contained many comments from the junior class members, including instances when the staff and senior class members deferred to them as subject matter experts. This collegial interaction continued during the noontime lunch periods when, in good weather, students would gather on the lawn outside to chat or play "kitten ball" (softball) against teams from the Naval Torpedo Station on Goat Island or the Naval Training Station, Newport.[50]

The demands of the curriculum were common to all students, regardless of their rank or service. They followed a schedule of classes Monday through Saturday from 9 AM until 4 PM, with their Wednesday and Saturday afternoons free.[51] The classes were long, the reading lists extensive, and the writing assignments numerous.[52] Instructors expected students to submit thesis papers on international law, command, and other subjects. While the theses were not graded (a feature that characterized War College classes until 1984), students who later wrote of these writing assignments did not remember them fondly, and historians have assessed most of the papers as simple regurgitations of class material. The papers do, however, document student attitudes toward the disarmament agreements (universally seen as major concessions to both England and Japan), geopolitics (surprisingly pragmatic views of the viability of defending distant overseas territories), and racial aspects of their potential opponents (dishearteningly dismissive of East Asian and eastern European cultures).[53]

Students conducted their individual study and committee work in office spaces assigned to them on the third floor of Luce and later Pringle Hall.[54] These spaces had to be vacated during the major wargames, when students playing the roles of the various component commanders were assigned to occupy the rooms. Prior to each major maneuver, the college staff wrote out "Detail"

The Players

memoranda, which listed players by name and position on one of the two opposing sides, students assigned to assist the maneuver staff with scoring or communications, and the rooms where they would work. The staff generally scheduled maneuver periods during the morning and left afternoons open for lectures or the review of staff solutions.[55]

War College instructors continually emphasized that the maneuvers were not contests but learning tools. The gaming manual *Conduct of Maneuver* emphasized that game directors "should impress upon the players that the exercise is a maneuver conducted for the purpose of training in strategy, order writing, and the exercise of command; and that it is not a game conducted for the purpose of seeing who wins or loses."[56] The companion manual *The Chart and Board Maneuvers* reinforced this philosophy, stating, "At the board, truth holds supreme rank," and reminded students that they should accept the occasional setback on the board as a learning opportunity and thank their adversaries for the lesson.[57] Despite these admonishments and Admiral Kalbfus's later attempts to ban the term "wargame" altogether, the students could not avoid seeing the games as a competition and they played to win, regardless of which side they found themselves assigned to.[58] Both Carl Moore '35 and civilian draftsman Philip Gaudet recalled catching some students in the act of surreptitiously trying to achieve unfair advantages on the maneuver boards.[59]

For most students, life outside the classroom in Newport could be very pleasant. There were no quarters provided for students, so all of them had to find accommodations in the town.[60] Earlier in the school's history local landlords had taken advantage of these temporary navy residents by charging high rents, but the college responded by working with the Newport Chamber of Commerce to investigate and resolve cases of price gouging. The school library, in an annex to Luce Hall, housed an ample supply of references and other reading material. The navy provided medical, commissary, and laundry services to the homes where the students lived.[61] Classmates from Naval Academy days were reunited and old friendships renewed. Carl Moore recalled,

"During the summer it was golf, swimming at Third Beach in Newport, and in the winter coasting and ice-skating and informal visits, short trips, get-togethers of all sorts. . . . We devoted our evenings almost entirely to working and studying. Our weekends were pretty free. Our families all had fun. Our children all had friends there."[62] The diary of Commander Harry L. Pence '25 contains only one entry that covers his time as a student, but the entry is three pages long and is a chronicle of social gatherings, bridge games, motor trips, and shopping visits to Providence and Boston.[63] Despite the Prohibition laws against the sale of alcohol, cocktail parties were frequent.[64]

While many former students and staff remembered their months at Newport fondly, the perspective of the War College student was much different from that of the Gilded Age millionaires or the modern sailing aficionados who have given Newport so much of its contemporary reputation. First, the school curriculum started in July and ended in May of the following year, so the students did not have the opportunity to take complete advantage of the glorious Newport summers. On the other hand, they did have the opportunity to experience the full fury of Newport winters. Thomas Hart's diary and Catherine Nimitz's memoirs are dotted with descriptions of extreme weather, concerns about obtaining sufficient supplies of coal for heating, and frequent illnesses of their young children.[65] Student life could also vary somewhat, depending on the seniority and financial resources of the officers. Those like Hart who were more senior, married, and better off financially had more opportunities to socialize and take advantage of what the area had to offer than junior, unmarried students.

Admiral Sims's dicta regarding no official navy duties notwithstanding, there were several opportunities for students to maintain contact with their operational brethren. Narragansett Bay was a frequent destination for the ships of the Scouting Force (the interwar designation for the cruisers and destroyers assigned to the East Coast), and students took advantage of those occasions to board visiting ships and renew acquaintances.[66] The Naval Torpedo Station across the bay from the college was also a previous—or future—assignment for many students and staff.[67] Many of the War College guest lecturers were visitors from one of the naval bureaus or the General Board in Washington.[68] Students were also keenly interested in their next assignments, and the more senior among them positioned themselves for command billets aboard cruisers or battleships. Assignments after the school varied, but the Bureau of Navigation made a War College degree a prerequisite for assignment to OP-12. Consequently, while students had no official military duties during their time at the War College, they were not necessarily isolated from their service.

Personal and official records of the period indicate that the War College leadership, faculty, and students all contributed toward creating a climate that encouraged experimentation and learning in a group setting. The primary contribution of the college itself, through its leaders, was to establish a formal mission of training, sell it to the greater navy, and then separate the students from their official duties and allow them to concentrate on their education. The War College staff largely removed or subdued the distinctions of rank and provided for frequent refreshment of the staff both from within and from outside the navy. They recognized the students as those who would influence future navy decisions on strategy and tactics, and they selected some truly original thinkers from the student ranks, especially those who were skilled at the give-and-take of wargame critiques, to serve as faculty.[69] The faculty members themselves constantly updated the wargame settings and lectures. They challenged the students without berating them, and they were generally flexible enough to adapt teaching material and methods when the students made a case for it.[70]

While demographically the students were certainly a homogenous group, their own writings indicate that there was nothing to prevent them from exercising their imaginations and trying out new ideas during the wargames. They were well educated, technologically savvy, competitive, and comparatively worldly. While few had seen any sort of combat during the world war, they were well familiar with the tactics and engagements that their allies had experienced. Students came to the school with open minds or, in the case of students like Carl Moore, had their minds opened soon after they arrived.[71] At the War College, they were among peers and friends, and their primary navy schooling and shipboard experiences made them used to working collaboratively. Their environment was relatively comfortable and relaxed, especially as compared to their lives aboard ships. They had no official duties and were free to concentrate on their thesis topics, their readings, and their wargame assignments. The distractions of family life such as child care and sickness were present but no more or less so than during any other assignment.

Overall, the documentary records of the staff and students reflect that they generally perceived the War College as a challenging, worthwhile experience and they responded to it with genuine effort.[72] There is no evidence among any of the documents reviewed that the students took the course of instruction less than seriously. While no interwar period War College alumni who left memoirs of their time at Newport gave such unqualified endorsements as did Admiral Nimitz, none of them even hinted that they found their experience to be ritualistic or irrelevant, even with the benefit of hindsight. While it is true that the absence of Spector's vocabulary in these sources does not prove relevance by itself, it does add strength to the argument in favor of relevancy.

The staff, faculty, and students were all ready to learn and experiment with new strategy, tactics, and technology, and the War College environment was conducive to such activities. Most important, the college experience meshed well with the navy's rotational assignment system, allowing students to take their new skills out to the fleet and return as instructors with these skills tempered by operational experience. With a foundational understanding of the players' experience established, the games these players participated in can be examined with a deeper understanding.

2

The Game Process

While lectures and independent study provided the foundation of the War College educational experience, the primary method for practical instruction at the college was the wargame. These events provided students with a venue to develop their abilities to estimate situations and to write appropriate plans and orders, and to exercise these skills while using information on existing fleets (both friendly and opposing) in geopolitically significant areas.[1] The War College maneuvers did not try to replicate *physical* aspects of warfare such as fatigue and physical discomfort but instead emphasized *mental* aspects by simulating how a naval decision maker would receive information. Accordingly, a large part of the efficacy of wargaming as a learning tool at the college depended on how realistically the game could replicate conditions that influenced player thought processes for a given class, and to what extent players could adapt the game to fit the learning objectives of a series of classes as world conditions evolved.

If one agrees with the judgments of previous histories, it is here in realism and adaptability of the wargame process that one would expect to find evidence of rigid structure, arbitrary judging, lack of flexibility, and resistance to change. In reality, what the game records show is a process that was anything but rigid and that actually became more accommodating to experimentation as the interwar period progressed. These wargames were not like chess, whose rules, pieces, and boards remained unchanged since the game's creation. The game board at the War

College was never the same, the players rotated in position and side, and the administrators and rules steadily evolved. Game planners on the faculty had to provide sufficient realism and relevance in the short term for one class but also maintain those attributes over the long term for future classes.

A navy fights wars at sea, not on boards. As with any simulation, there are limits as to how much accuracy or realism developers can inject into a wargame, and every one of these limits in turn provides an opportunity to decrease the game's relevance. Distilling something as complex as armed conflict into a game that features learning objectives, rules, and scores requires that many, if not most, of the complexities be artificially managed. Such artificialities, which take the form of conditions, limitations, and assumptions, are present in all games to some extent to keep the game focused and manageable. For example, a game compresses time into fixed periods or turns. Dice, spinners, or other mechanical means inject randomness. Referees or rules direct transgressions outside the physical and imaginary boundaries of the game back inside. The game itself injects biases through aspects such as opposing objectives, scoring of results, referee experience, and whether the game proceeds in real or compressed time. Preconceptions, prejudices, artificial boundaries, and limited scope can sometimes hinder original thinking on the part of the players. Often the line between game and reality blurs, and those artificialities can affect the way the reality is "played" out.[2] The challenge to wargame developers is to strike the right balance, to provide sufficient structure but at the same time prevent the game from devolving into a free-for-all.

In a wargame the scenario and game rules represent this line between game and reality. The documentation of Naval War College wargame scenarios and rules, the official papers of the staff that orchestrated the games, and the unofficial papers of the students that played them show how the staff managed their way around the inherent biases to develop the interwar period games into a realistic and adaptable instrument for training and experimentation.

"Scenario" is a common wargaming term for the game's context

The Game Process

or setting, and it is probably the most obvious gauge of a game's realism. The scenario describes such attributes as the opponents, their respective objectives, orders of battle, the physical geography of the battle space, and the geopolitical background that brought about the conflict.[3] A scenario may represent the real world, or it can be an imaginary one constructed to emphasize certain situations or learning objectives. Today as well as in the past, wargame scenario developers walk a fine line to achieve a proper balance of realism and educational relevance. Their conundrum is that the most realistic and detailed scenarios produce results and lessons that are only narrowly applicable. But the broader and more high-level a scenario, the less concrete information can be drawn from it to guide player actions.

During the interwar period, wargame developers at the War College based their scenarios on the color-coded war plans that detailed courses of action to take if the United States should enter into hostilities against another nation.[4] During the first half of the interwar period there were a number of these contingency plans, each focused on a single opponent.[5] The two that received the most attention from U.S. military leadership were War Plan ORANGE, the plan for conflict with Japan in the Pacific, and War Plan RED, which featured England as the opponent in an Atlantic setting.

The recurring play of RED—an ally just a few years previous—as the opponent in War College wargames may appear singular at least. Michael Vlahos interpreted RED games as indicative of how the U.S. Navy measured itself against the Royal Navy.[6] While lecture subjects and student diaries reflect awareness of the volatile situation in postwar Europe, none of these documents provide any indication that naval officers of the time ever imagined that the United States might go to war with England.[7] However, a conflict of some sort in the Atlantic area was not beyond the realm of the possible. Wargame scenarios have been, and are still, selected to stress certain conditions or achieve specific objectives. If one looks beyond the color codes to the college learning objectives, one sees a series of Atlantic and Pacific theater game scenarios. RED games allowed students to become familiar with the geography of different regions such as the Caribbean and the South Atlantic. They also provided plausible if not completely realistic scenarios for defensive situations, or ones where BLUE could find itself at a numerical disadvantage.

While the BLUE versus RED situations might have been somewhat contrived, the same cannot be said of the BLUE versus ORANGE scenarios. War College staff and OP-12 war planners considered ORANGE the most likely of all the war plans to be executed.[8] Accordingly, the majority of the War College maneuvers throughout the interwar period were set against an ORANGE backdrop.

The climax of each class was a large-scale multiweek affair variously referred to as the "Big Game," the "Major Operations Problem," or the "Large Game." ORANGE was always the opponent in these large-scale games, and the strategy and tactics staffs aligned the scenario to reflect specific phases of the current version of War Plan ORANGE. The Big Game, generally a series of maneuvers played in stages set in the same scenario, started with chart maneuvers for logistics planning, search, and screening, and it culminated in a large-scale tactical board maneuver. This game occurred in the spring toward the end of the course, though it was not always the final game played.

The navy's vision for the war plan version of ORANGE may have been simplistic in the first half of the interwar period, but War College instructors did not generally embellish or otherwise modify the ORANGE wargame scenarios or stray from the realpolitik. Orders of battle and geography were as close to real world as possible.[9] The staff and faculty members rigorously researched physical ship characteristics (armament, speed, resistance to damage, endurance), and students played those ships as realistically as they knew how. The ORANGE side capabilities were not inflated or handicapped, and this was one of the more significant aspects of War College games. Michael Vlahos cited student thesis papers as evidence that student officers viewed their Japanese foes with a trace of imperialist disdain.[10] Diaries and correspondence written by the students bear this out to an extent, but in hindsight one must balance Vlahos's judgment against the fact that the ORANGE was the opponent in all of the Big Games until 1941, and that there was no permanent opposition team made up of faculty members. Players on the ORANGE team were student officers who just happened to find themselves assigned to that role.[11] War College instructors evaluated student teams and their respective results by the same criteria regardless of their side or position, that is, by how accurate their estimates of the situation were, how well their decisions capitalized on the strengths of their own forces and the weaknesses of their enemy, and how well they communicated their intentions to their fleets. The impact of this construct is that students on both sides played their assigned roles to the best of their abilities.

War College classes played out scenarios of the interwar period in two broad classes of maneuver: chart and board. Chart maneuvers encompassed "Search & Screening" and "Strategic" games (designated Strategic, STRAT, or simply S games in the War College taxonomy). These maneuvers exercised student ability to move fleet units toward an objective, to conduct scouting and screening operations, and to maintain the necessary flow of supplies. The chart maneuver process started when students received a paper copy of the problem statement from the strategy or tactics instructor. The students would be expected to develop an estimate of their respective situations and an appropriate solution either individually or later on in small "committees." They turned these solutions in to the staff, which picked one of them for the whole class to play.[12] Players from each team based their development of movement orders on their understanding of their respective situations, and draftsmen translated these into plots on large-scale nautical charts. Once the opposing fleets came into contact, the chart maneuver transitioned to a board maneuver or "Tactical" game (designated Tactical, TAC, or T games) conducted with ship models on room-sized gaming boards in Luce Hall.[13]

According to the schedule and class records of the period, War College senior classes played twelve to thirteen games each.[14] The first of these were demonstrations of World War I battles to illustrate the wargame concept. Previous historians have misinterpreted this practice of using Jutland, Coronel, and the Falklands for practice games as actually playing those situations as real games. Ronald Spector was correct in stating that the War College played the Battle of Jutland every year, but the purpose in playing it was simply to demonstrate to the students how a real engagement would look when simulated on the game board, and not as an actual maneuver. In these demonstration games, students moved their ships according to a script—the only real "script" in the wargaming process—based on the actual battle, provided a narrative of the move, and then discussed its significance to the overall battle. In this way, re-creating the historical battle on the game board was in the same vein as the Gettysburg staff ride at the Army War College. Demonstration wargames were followed by basic strategic and tactical maneuvers designed to have students practice development of their own estimates of the situation on a simple scale. As the class continued, these maneuvers expanded in scope, length, and complexity.

Table 1 illustrates a game schedule for the class of 1930, with three demonstration problems, two historical situations acted out on the board, and two weeks of "quick-decision" exercises. Starting with the class of 1929, the Department of Operations initiated quick-decision maneuvers to simulate situations where decision makers needed to develop estimates and orders in a compressed timeframe and usually under some duress. Quick-decision maneuvers were set up on the game board behind screens or curtains. On a signal, instructors removed the screens to reveal the situations to the students, then started the clock and timed the students while they developed their situation estimates, decisions, and orders. Once the students turned their orders in to the instructors, the maneuver proceeded in accordance with the student decisions.[15]

After the quick-decision phase, the classes played a series of problems culminating in a two-month-long Big Game. The maneuver "outline" for the operations problems offers no script for operational objectives but only states the educational objectives of developing estimates of the situation and developing orders in accordance with those estimates. Also, the days scheduled for the maneuvers run from one to two weeks for the demonstration maneuvers to fifty and sixty days for Operations Problems III and IV.[16] As was the case in many years, the Big Game was not the final maneuver problem. The staff generally scheduled one more strategic or joint maneuver that took up the last weeks of the class and probably allowed the faculty sufficient time to process and critique the Big Game results before graduation.

From the perspective of World War II naval combat, a major gap in the maneuver spectrum was that almost all the large-scale games culminated in a fleet action. The wargame catalog for strategic, tactical, and operations maneuver scenarios included convoy protection and amphibious assault situations, but these were generally backdrops to fleet-versus-fleet battles. Beyond the few days set aside for minor and quick-decision tactical problems, there were no small-scale engagements.

In any wargame, the conflicts between realism and artificialities are negotiated within the game rules. Game rules, or at least how often game developers revisit and change them, are also a measure of a game's adaptability. Successive versions of War College game rule publications reflect the staff's efforts to maintain a balance of allowing freedom of action without being overly prescriptive. Additionally, these publications record how the staff accommodated changes in doctrine and technology.[17] Five of these documents, *Maneuver Rules*, *The Naval Battle*, *The Conduct of Maneuvers*, *The Chart and Board Maneuver*, and the fleet description booklets, first appeared in the initial years of the early phase and were continually updated and used afterward. One can consider the *Estimate of the Situation* pamphlet to be part of this group, but it did not prescribe any game processes. The five manuals described to staff and faculty how to execute the games, and each had a different lineage.

Maneuver Rules was a detailed manual of wargaming data prepared by the Department of Operations.[18] It contained all the game tables and mathematical functions that today are hosted in a computer as algorithms. These included factors such as ship fuel consumption, weather effects on visibility and communications, ship and aircraft performance, and the all-important gunfire

Table 1. Senior class of 1930, course outline

Period	Days	Subject	Outline	Duration (active days)
1	1–3 July	Organization	Opening address, assignment to rooms, general preliminary preparations	3
2	5–10 July	Service of Information and Security	Presentation of a search problem	5
3	11–13 July	Command	Command—standard publications	3
4	15–20 July	Demonstrative chart maneuver	Presentation of the "Maneuver Rules" and "The Conduct of Maneuvers"; Exercise in the conduct of the chart maneuver	6
5	22–27 July	Demonstrative board maneuver	Presentation of tactical features of the "Maneuver Rules"; Exercise in the conduct of the board maneuver	6
6	29 July–14 August	Demonstrative operation problem	Presentation of a demonstrative operation problem, with a view to the student estimating the situation, formulating his decision, operation plan, and the resulting tactical decision and plan	15
7	13–21 August	Coronel and Falkland Islands (with junior class)	The detailed movements of the German Far Eastern Squadron and the Allied Forces in search of them will be traced from 27 July 1914 up to and including the destruction of the *Dresden*	6
8	22 August–7 September	Battle of Jutland (with junior class)	Maneuver on the maneuver board with a narrative of each move by the students assigned as unit commanders	14
9	9–22 September	Minor and quick decision tactical problems	Specific tactical situations of limited scope requiring tactical estimates and decisions, and the illustration on the maneuver board of the effect of such decisions	12

10	23 September-12 October	Operations Problem I	An estimate of the situation, with resulting decision, plans, and operation orders; based on a joint mission, joint decision, and joint plan	18
11	14 October-2 November	Tactical Problem I	A tactical situation, based upon a strategic situation, imposing a fleet engagement upon both sides	17
12	4 November-7 December	Operations Problem II	An estimate of the situation, with resulting decision, operation plan, and orders; based upon the operation orders of higher authority	30
13	9 December-15 February	Operations Problem III (with junior class, 3 January to 15 February)	An estimate of the situation, with resulting decision, operation plan, and orders; based upon a concept of the war, basic mission, basic decision and campaign plan	50
14	17-21 February	Joint operations	Presentation of Op. VI 1929, "Joint Action of Army and Navy" and "Joint Operations—Landings in Force"	5
15	24 February-3 May	Operations Problem IV (with junior class, 16 March to 3 May)	An estimate of the situation, with resulting decision, operation plan, and orders; based upon a concept of the war, basic mission, and basic decision	60
16	5-24 May	Strategic Problem V	An estimate of the situation, with resulting basic decision, including decision as to the deployment of the fleet and its initial employment: based on a concept of the war, basic mission, and initial task	15

A table showing the outline of course provided to the students of the Naval War College senior class of 1930. *Source:* U.S. Naval War College, "Outline for Course: Senior and Junior Classes of 1930," Folder 1511, Box 50, RG4, Publications, 1915–77, NHC.

and weapon effects tables.[19] Game umpires referred to *Maneuver Rules* to determine when ships could detect their targets, when they were within gun range, how many hits they would achieve in a given salvo, and what the effects of those hits on the target ship would be, such as a reduction in speed or ability to return fire. Many of the later conflicts regarding the effectiveness of air attacks on surface ships were negotiated through changes to the tables in this document. The first edition of *Maneuver Rules* appeared in 1922, and the manual was updated in 1929, 1935, 1937, and 1941.

The Naval Battle came to the War College by a less direct path than the other guidebooks. Then-captain Harris Laning '22 originally wrote this manual as his tactics thesis assignment. *The Naval Battle* was essentially a tactical how-to book for aspiring fleet commanders. In it Laning described the roles of each ship and aircraft type, basic tactical principles, the influence of environmental conditions, and how to conduct operations such as advancing, deploying for action, and scouting. Laning followed Sims's lead by comparing the fleet to a football team, where the team's success depended on each player dutifully performing their designated role.[20] Laning's thesis apparently impressed the War College faculty sufficiently that they decided to incorporate it as a pamphlet to aid students in the execution of their tactical games. When Laning returned to the War College as an instructor in 1923, he was surprised to see his thesis had become part of the curriculum, but as Tactics Department chair, he greatly expanded its use.[21] While in some ways *The Naval Battle* placed boundaries on student-officer horizons by defining the limits of each ship's role, it also expanded the horizons of officers whose pre–War College experiences were limited to one or two classes of ship. As was the case with *Maneuver Rules*, the Tactics Department periodically updated *The Naval Battle*, and it remained in use at the War College until the mid-1930s.

The next guiding documents, *The Chart and Board Maneuver* and the very similar *Conduct of Maneuver*, were detailed descriptions of the wargame process. Judging from the language used in these manuals, the first was probably intended for students playing the roles of opposing fleet elements and the second was probably written for the maneuver detail. Both of these manuals described the physical setup of the game rooms, how the games were officiated, roles of the support staff, what equipment and information would be available to each team, how each move was to be executed, how the players and maneuver staff communicated, and how results would ultimately be decided. First published in 1928 and regularly updated throughout the interwar period, these documents provide the most complete description of the wargame experience during the interwar period. For the purposes of this book, *The Chart and Board Maneuver* and *Conduct of Maneuver* are especially significant for what they do *not* contain, which are scripts or other prescriptive directions regarding game moves. These rule books placed most of the discretionary power for executing the moves into the hands of the students, and for running the maneuver process into the hands of the game director.

Fleet description booklets written by the Department of Intelligence made up part of the game outfit.[22] The information in these booklets represented the latest intelligence regarding U.S. and foreign navy ship, submarine, and aircraft types; their armament, endurance, and speed; and other specifications relating to their fighting capabilities. The War College staff made every effort to obtain the latest specification data to represent the naval assets played in the wargames accurately. This data was not always completely accurate or truly representative, but the material provided to the wargame players was the most current data available.[23] In this respect, the fleet description booklet exemplifies much of the information available to War College students: up to date when written, if not exceptionally forward-looking.

Perhaps the most distinctive aspect of the interwar period wargames at the Naval War College was the physical setting. While the arrangements and equipment in the maneuver rooms in Luce and Pringle Halls might seem quaint today, they represented the state of the art for their time. Luce Hall opened in 1892, with lecture rooms and miscellaneous services on its first floor and student and staff offices on the second and third floors.[24]

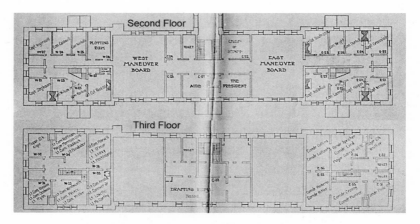

FIG. 5. A floor plan of the second and third floors of Luce Hall, circa 1928. The east and west maneuver board rooms are indicated on the second floor, while offices of the staff and senior students are shown by name. "Floor Plan of Luce Hall and Mahan Hall," Folder 1270, Box 7, RG4, Publications, 1915-77, NHC.

The most prominent features of Luce Hall were two sixteen-by-twenty-five-foot atrium style Maneuver Board Rooms in the east and west wings. Classes conducted their tactical or board maneuvers in these rooms. Figure 5 shows these rooms, along with staff offices and student study rooms on the second and third floors of the east and west wings. The names written in the offices on the floor plan are office assignments that correspond to the students and staff assigned to the War College class of 1928.

The hallmark of the maneuver rooms and the associated equipment was their simplicity, which in turn made them extremely flexible. Large gridded game boards were the largest features in each of the rooms. These were first set up first on wooden sawhorses and later laid directly on the floor to provide maximum gaming space. The grids on the Luce Hall boards were forty inches square and were subdivided into smaller four-inch squares. The scale, or the ratio of distance on the game board to the corresponding distance on the imaginary ocean surface, for the Luce Hall boards could be varied but was typically set at 250 yards to the inch. Instructors drew chalk lines on the boards to delineate landmasses, and students used small-scale models to represent

The Game Process

the individual ships in their fleet. Different colors on these models represented the different ship types. The ship types included all current navy ships as well as some imaginary classes designated as auxiliary (with an "x" appended to their standard navy designation, such as XCV for a merchant ship converted to an aircraft carrier) and second line (appended with an "o," such as OCL to designate an older light cruiser used for fire support tasks).[25] Auxiliary and second line designations were a convenient way for the staff to add additional capability to either side when needed to support a given mission. Players moved these models across the boards in accordance with orders received from their respective commanders and in keeping with the capabilities of the ships represented. Figure 6 shows a scale-model game board on display in the Naval War College Museum. In this photograph, the gridded board represents the game room floor. The opposing forces (RED to the left and BLUE to the right) are approaching each other at an angle in columns. Model ship formations are on the tracks. The small cards with arrows laid next to the ship tracks indicate gunfire. The large white angle originating from the RED column in the center of the board represents a spread of torpedoes fired ahead of the three BLUE columns approaching from the top of the photograph. This collection of models, cards, and lines represents a RED force gun and torpedo attack on BLUE, followed by an evasive turn away. The lead BLUE column is turning toward RED to pursue, while the trailing BLUE columns are returning RED fire.

Physical separation in different wings of the buildings blocked opposing teams from seeing each other, and the game director maintained a separate master plot (which was off-limits to student players) in one of the maneuver rooms where he could see all movements and make rulings.[26] Figure 7 shows a game in progress in Luce Hall during 1906. Labeled columns on the chalkboard in the background reveal this as a RED versus BLUE game. Students in civilian clothes are moving model ships on an elevated game board, literally "in the game," in contrast with the 1895 illustration of the chart maneuver in the introduction chapter. Game directors and umpires are standing behind the elevated railing visible on the left.

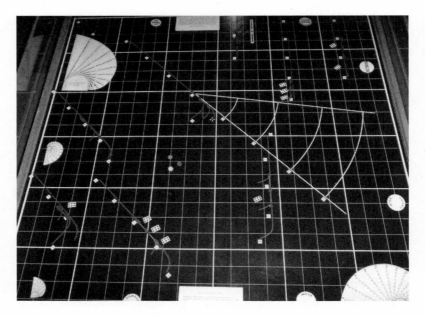

FIG. 6. A scale model of a Naval War College game board as exhibited in the Naval War College Museum. The model shows a tactical game as well as turning circles, dice, and badges actually used in the games. Courtesy of the author.

In an important way, the gridded maneuver boards in Luce Hall defined the world of naval combat by reflecting the supremacy of the battleship. In 1928 the maneuver board measured sixteen feet, eight inches, by twenty-five feet, eight inches, with a standard scale of 250 yards to the inch.[27] A quick bit of math will reveal that the scale maritime battle space for these games was only twenty-four by thirty-eight nautical miles. Even the slowest aircraft of the early phase could traverse the dimensions represented by this board in less than thirty minutes. However, naval commanders of this era did not consider the aircraft to be a primary tactical weapon due to their small weapons payloads. Naval tactics, and the game board on which they were exercised, were centered on the large-caliber gun. The maximum ballistic range of the 14″/50-caliber guns of the *New Mexico* class battleships were about twenty-four nautical miles, though under battle conditions the effective range depended on the ship's ability

　　　　　　　　　　　　　　　　　　　　　　　　The Game Process

FIG. 7. A wargame in progress in Luce Hall, circa 1906. The students are wearing civilian clothes as opposed to navy uniforms, and the game board is set up on sawhorses. In later years the board would expand in size and be placed directly on the floor of the maneuver room. Courtesy of the Naval War College Museum Collection.

to mark its intended target visually. The tactical game boards in Luce Hall were metaphorical boxing rings for bouts between naval heavyweights.

Figure 8 shows a wider view of the west maneuver board room with a game in progress during the late 1920s. This photo was part of a report that documented how the physical dimensions of the rooms limited the scale of scenarios that the college could emulate.[28] The game board has been moved from its former position atop sawhorses to occupy most of the maneuver room floor, which allowed interwar period students and staff even greater access to the game than their 1906 predecessors had. As in figure 7, these students are wearing civilian clothes (with the exception of one army officer on the right). A number of students are kneeling on the floor, making and marking ship movements. The lines strung over the students' heads are where curtains for quick-decision problems were hung. The uniformed marine standing next to the chalkboard is one of the messengers who carried written orders to and from the rooms where students playing the roles of staffs were isolated.

Several items of game equipment are visible. Two of the students are using turning cards to calculate the sea space required for a ship steaming at a given speed to make a turn. There are

FIG. 8. The west maneuver board room in Luce Hall during the interwar period, showing a wargame in progress. Courtesy of the Naval War College Museum Collection.

three long banded poles visible in the picture, one leaning against the left side of the chalkboard, one in the back center of the room, and one held by the student kneeling in the center of the floor. These are range wands, used to measure the distances between individual ships on the board.[29]

In the early 1930s the War College underwent a major expansion. By this time the size of the senior class grew to forty, and the junior class—who also participated in wargames—had grown to thirty-seven students. Pringle Hall, a new gaming facility adjacent to Luce Hall, was constructed to answer what must have been a critical need for maneuver space. Pringle Hall was a three-story building with 40,990 square feet of floor space, and its major feature was a sixty-eight-by-ninety-two-foot maneuver room on the second floor surrounded by an observation mezzanine above.[30] The scale for this maneuver board was larger—six inches to a thousand yards—than the Luce Hall boards.[31] What this meant was that the maximum space available for a game

The Game Process

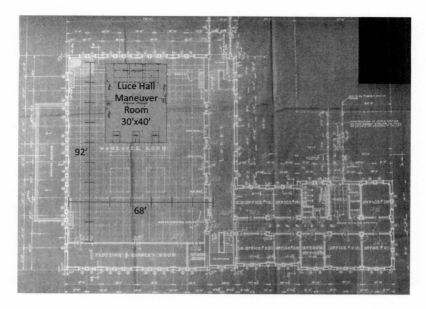

FIG. 9. The plan of the second floor of Pringle Hall, showing the
increase in game area (over six thousand square feet) as compared with
the maneuver room in Luce Hall (twelve hundred square feet). Navy
Department Bureau of Yards and Docks, drawing 114764, "Naval War
College Newport RI, Extension of Main Building; Second Floor Plan,"
Folder 9, Drawer 5, Folio III, NHC.

increased by over five times the area, from just over nine hundred scale square nautical miles to over six thousand. In terms of
scenarios, this increase equated to the ability to simulate much
larger opposing fleets maneuvering over the visual horizon, still
in contact by aircraft but not visible to each other on the surface.
Figure 9 shows the floor plan for Pringle Hall with the plan for
one of the Luce Hall maneuver rooms superimposed to illustrate
the increase in gaming space.

Figure 10 shows the interior of Pringle Hall shortly after its
opening in 1934. The maneuver board grid is printed on the
linoleum, which stretches from wall to wall as opposed to being
marked off on wooden boards. The game assistant standing in the
center rear of the photo and the janitor in the lower left provide a
gauge for the size of this room relative to those in Luce Hall. Ship
models, range wands, and turning cards lie on the floor inside

U.S. NAVAL WAR COLLEGE, NEWPORT, R.I.   8 May 1934.
NWA Reqn. 117.   View looking west in Maneuver Room,
showing cash carrier system.   No. NTS. "C" 34-129.

FIG. 10. A view of the game floor in Pringle Hall shortly after its opening, showing the observer's mezzanine and some game equipment in place. Courtesy of the Naval War College Museum Collection.

the rope barrier. The pneumatic tube ("cash carrier") communication system, which replaced the marine messengers, is visible on the back wall. Curtains are partially deployed around the back left corner of the board. Pringle Hall remained the primary maneuver room until 1957, when the War College replaced the manual game boards with the Naval Electronic Warfare Simulator (NEWS), a computer-based game system.

Back in Luce Hall, builders installed floors on the third-floor level above each of the old maneuver rooms to provide more space for staff and student offices, but the War College retained a "junior game board" maneuver room in the new space above the former east maneuver room.[32] Figures 11 and 12, respectively, show east- and west-facing views of that space. Together, these photos capture a Junior class wargame in midstride with some of the gaming arrangement shown to better advantage than in the Pringle Hall picture. The photo is undated, but the candlestick telephone at the back of the room between the doors in figure 11 points to a time during the 1930s. The folding screens

FIG. 11. A composite of two photographs of the third-level maneuver room in Luce Hall, facing west. After the opening of Pringle Hall, this space was used by the junior class. Courtesy of the Naval War College Museum Collection.

stacked against the wall by the doors could be set up to prevent players from seeing certain areas of the board. Range wands are visible on the table in the left foreground and by the doors at the rear. The sign on the table at left rear is marked "C in C," indicating that the force commander operated from this position. Ship markers and game equipment are on the floor at lower right.

In figure 12, which shows the same room as figure 11 but with the photographer facing the opposite direction, the wall clock above the door indicates an afternoon scoring period when the players were most likely attending lectures. The chalkboard records game moves and weather conditions, and the sign on the table on the left indicates the post of the communications director. More folding screens are stacked against the far wall. In the morning of the following day, the students will return to their posts and resume the game where they left off.

At the start of a chart or board maneuver, maneuver staff and students gathered in various offices in the east and west wings of Luce Hall, with opposing teams physically separated from each other. Elements that represented different fleet units from the same team were further separated in different rooms if the ships they controlled were out of visual range of each other. These elements maintained their own plots and communicated via written messages carried from room to room by marine messengers.[33] The game director, who was most often the head of the Operations or Tactics Department, determined the progress of the

FIG. 12. A composite of two photographs of the third-level maneuver room, Luce Hall, facing east. Tables for staff members and game equipment are visible on the floor and along the walls. Courtesy of the Naval War College Museum Collection.

game. A staff of up to twenty-three officer assistants and a civilian cadre of draftsmen, stenographers, and recorders assisted him.[34] These included assistant directors for operations, moves, gunnery, and the like; move and communications umpires; gunfire scorers; and damage recorders.[35] Students assigned to the maneuver detail for the game filled most of these positions.[36] Throughout the maneuver, the director's task was to manage all functions of the wargame process. He oversaw the master plot where his assistants tracked both sides' movements, he established the weather conditions for the maneuver, and he controlled the game clock that determined how much time each team had to complete their moves. With his large staff and ability to dictate time and weather, it is small wonder that Michael Vlahos characterized the director's role as "Olympian."[37]

At the start of the game sequence, opposing fleet commanders reviewed intelligence reports, developed their initial estimates of the situation, and generated orders to their subordinate units. Students on the commanders' staffs translated the orders into messages, which they passed on through the game director. The director determined how much of the message would be passed to its recipient and how long it would take to be transmitted.[38] The director also set the game clock for moves in three- to fifteen-minute increments (meaning that each turn would take

The Game Process

that much game time).[39] Students plotted their movements on charts during a chart maneuver or on the board during the tactical phase.[40] Moves on the board for large forces could be very complicated, such as maneuvering large formations, launching aircraft, or laying mines. If a fleet commander decided to lay a smoke screen, the game staff would place small screens of paper or cloth across portions of the board to represent the smoke.[41] Students plotted maneuvers precisely by using turning circles. Those playing the roles of aviators and submariners developed their estimates and orders in separate rooms and were not permitted to see the game board unless they were performing reconnaissance functions. On those occasions, the game director allowed them only a three- to six-second glimpse.[42] When a scouting unit (in a plane, ship, or submarine) sighted the opposing force, they passed the appropriate information via messenger back to their commanders. When units engaged, the maneuver staff consulted fire effects and ship damage tables to determine the results.

Horn blasts (appropriate for a student body used to ship whistles and sirens) would signal the end of the move periods and time for students to turn in their moves.[43] At the end of each move, staff recorders documented all pertinent information. After the morning gaming period ended, stenographers recorded ship movements and their final positions. Draftsmen transcribed these movements as tracks on blueprints for review the following day, and for the critique that followed each game. This cycle of planning, movement, calculation of results, and recording continued until the director decided that the class had achieved the learning objectives for that session, at which time he concluded the game.[44]

After the staff compiled and reviewed the game results, students assembled for a group discussion of what happened. Faculty members who played the roles of umpires and game staff conducted these discussions and prepared critiques of the student teams. The depth and detail of these critiques depended a great deal on the critique writer. Harris Laning wrote extremely detailed critiques with move-by-move narrations and comments. Others writers deferred move descriptions to other documents and kept their comments at a higher level. The ensuing discussions at the critique sessions generally included a great deal of comment and occasional arguments. Staff recorders captured these and included them as part of the maneuver records.[45]

With all the necessary planning, communications, plotting, and calculations, the actual time for a move that represented three to fifteen minutes of game time could be very long. Figure 13 illustrates how the calendar days ("game days") spent conducting the Big Games increased between 1927 and 1933, while the days actually simulated ("war days") decreased over the same period. This is one reflection of how the games increased in complexity and generated larger volumes of data as the interwar period progressed. Wastage of time during the game was of particular concern to Carl Moore during his tour as an instructor from 1936 to 1937, and he instituted changes to reduce students' idle time between movements. Moore consolidated and simplified the scoring forms, and chased off students who had been carrying on extraneous conversations, playing cards in the maneuver room between moves, and otherwise distracting the maneuver staff.[46]

Rather than reflecting an irrelevant and rigid ritual, primary source material on the interwar period wargaming process shows the complete opposite. These sources document wargame scenarios based on reality, flexible procedures that provided a balance of control and freedom of action, and the mechanisms necessary to adapt to new developments.

While some critics today might judge the game process as unrealistic in its ability to simulate a naval combat environment, the game setup did mirror some of the physical experience of a fleet staff. The rooms did not pitch and roll and there was no gunfire to contend with, but the important aspect of having to deal with partial, heavily processed, and time-late information was quite accurately replicated. Unless the flagship itself came into contact with the enemy, the staff would be isolated in a shipboard plotting room, relying on radio messages, flag signals, lookout reports, and plots made on large-scale charts to build and maintain their awareness of the situation. Intelligence regarding opponent loca-

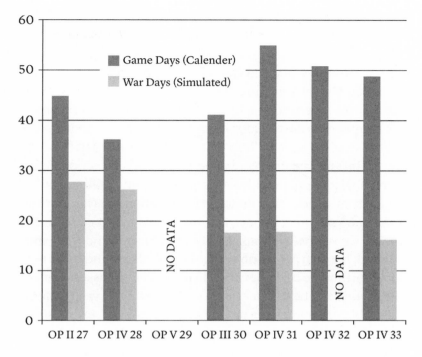

FIG. 13. Damage absorbed by BLUE and ORANGE ships during the TAC IV-35, between moves 4 and 30. This chart illustrates the number of calendar and game days spent on the trans-Pacific games during the years 1927 to 1933. Compiled by the author using data extracted from the Research Department report "Analysis of Trans-Pacific Problems as played at Naval War College, Newport," July 1933.

tion and disposition was subject to delays imposed by the director. This was intentional, as the staff believed that "the strain imposed by long periods of time passing without information is an element of actual war which should not be disregarded."[47]

Thomas Hart provided an endorsement of wargame relevancy and realism in his personal diary. Hart wrote daily entries, which provide a relatively unfiltered view of how students perceived the wargaming experience. Hart was not a cheerleader for the War College—his comments regarding most of the guest lecturers are anything but laudatory—but his views of the wargames are consistently positive. He recorded his impressions of game realism on 28 August 1922. "Having very interesting stunts at the college now—real fights on the game board and it is remarkable how closely we can simulate the real thing. Too bad the whole Navy doesn't do these games—it wouldn't guess so much if it did."[48] The date of the entry and the reference to "game board" indicates that Hart is referring to an early board maneuver, either Tactical II / TAC 92 or Tactical III / TAC 93 (both RED games). His comment is notable, for though he describes the wargame as a close simulation of "the real thing," like most of his classmates Hart had no firsthand experience in a fleet action. Social psychologists might call Hart's opinion "confirmation bias" (a tendency to search for or interpret information in a way that confirms one's preconceptions), and historians might call it a reinforcing set of observations. However, if Hart found the games to match his own perception of reality, he also found them to be both worthwhile and relevant. He reiterated his assessment on 1 September: "The 'Battle' was very interesting—I was badly but honorably defeated; lived up to the best traditions, etc. I can see that the War Game does give real training."[49] Hart makes two points in this later entry. First, he admits to a setback, reflecting that there were unambiguous outcomes to the wargames and that they were important to the students. Second, he asserts that the wargames provided relevant training that justified the effort put into them.

The other aspect of the game that endured was its adaptability to new strategies, tactics, and technologies. This flexibility is reflected in the game manuals and materials. While the bulk and scope of the manuals might seem overly prescriptive and thereby restrictive, one must bear in mind that the game rules left the strategy and tactics embedded in the moves to the discretion of the student-officers. Furthermore, the college needed some structure and organization to shepherd an average of sixty-nine senior and junior students through a progression of demonstration, strategic, quick-decision, and tactical maneuvers every year. The number of times these manuals were updated gives some measure of the War College's ability to accommodate change, or at least the staff's efforts to adapt the game process and incorporate new information as it became available. The game board and sequence were modified over time, but these features would remain the pacing aspects of the game in terms of geographic space and calendar time. It is no wonder then that the War College wargames were some of the earliest examples of implementing computer-based games on a large scale.

While the game's players—the staff, faculty, and students—came and went and each contributed to naval tactical and strategic development in the short term, the game was the long-term factor, providing a relatively stable environment for those short-term developments to build on each other. With the players and the overall gaming concept itself introduced, the next step in this examination is to review in some detail the progression of games conducted during the interwar period, and to assess how the game experience and results influenced the opinions and decisions of those players once they faced real combat in World War II.

3

The Early Phase, 1919–27

E xamining primary source material about the game players and process leads to a deeper examination of the games themselves. If, in a manner established in the first two chapters, one accepted the characterization of wargames of the interwar period as irrelevant, scripted rituals, one would expect to see little to no evolution of strategy, tactics, and thought in the game records. While the wargame records of the early phase years of 1919 to 1927 reflect a navy still coming to grips with technological and doctrinal developments of World War I, they show the point of departure and reflect the magnitude of the challenge facing the college and the navy as they prepared for the coming conflict.

What characterizes the greater navy of the early phase of the interwar period is decline—a dramatic regression from a position of power and prestige to a national afterthought in terms of funding and strength. During these years, War College students participated in wargames under the twin specters of capital ship scrapping and less-than-optimal force ratios compared with their opponents. At the same time, these imaginary campaigns were conducted against a backdrop of continual American engagement overseas in "Banana Wars" such as the occupations of Nicaragua, Haiti, and the Dominican Republic, and the continuing presence in China.[1] Conversely, War College records reflect that while the greater navy was enduring this period of reduction in strength and dispersion of effort, the college experienced both a time of growth in its size and influence in general and intense development of its wargaming program in particular.

One year after the Armistice ended World War I, the U.S. Navy had come very close to achieving the "second-to-none" vision established by the General Board just four years previously.[2] From a small organization that ranked a distant third in terms of worldwide capital ship tonnage in 1915, the navy grew to a close second place behind Great Britain by the end of the war.[3] But despite the imposing size of this force, the wartime U.S. Navy had seen little in the way of actual combat. The major fleet battle that senior navy officers had studied and practiced for most of their careers had already occurred at Jutland the year before U.S. entry in the war. The last time the U.S. Navy had conducted a major surface engagement was against an understrength Spanish fleet at Santiago de Cuba during the Spanish-American War twenty-two years prior. The wartime commander of U.S. naval forces in Europe, Admiral William S. Sims, served as naval attaché to France during the Spanish-American War and had no combat experience at all.[4] So while in terms of size the U.S. Navy was second only to the Royal Navy, at the start of the early phase it ranked behind the navies of Great Britain, Germany, France, Italy, Russia, and even Japan in terms of experience in battle.

The fortunes and status of the navy took a major downturn because of the combined impacts of the end of the world war and a new fervor for pacifism and disarmament, which rendered the navy's mismatch between size and experience largely moot. These trends had a variety of causes, some ideological and some pragmatic. The overall ideological cause was popular disenchantment with war as a vehicle for furthering national interests. This feeling was particularly strong among the combatant nations who had lost so many of their people and so much of their treasure during the "war to end all wars." On the pragmatic side, there was a strong desire on the part of leaders of the western colonial powers (which in 1919 included the United States) to counter the rise of Japan and maintain the naval primacy that was so important to the continuing security of their overseas possessions. Leaders of the major powers saw the pacifist movement as a convenient rationale to avoid an expensive naval arms race while maintaining the prewar status quo. Enthusiasm for disarmament both in America and in the rest of the western world manifested itself in the 1920s in a number of formal organizations and agreements, culminating with the Washington Naval Conference of 1922.[5]

While naval officers of the early phase fulminated over the abstracts of the Five Power Treaty and dealt with small-scale incidents and flare-ups in real life, they measured their combat prowess on the game board against a yardstick set by former allies Great Britain and Japan. Strategically, bringing the naval opponent to a climactic Jutland-like fleet engagement remained the centerpiece of American naval doctrine. Alfred Thayer Mahan died in 1914, but under his enduring influence, naval war plans, fleet exercises, and wargames of the early phase all featured common stages to enable this major fleet engagement. The Mahanian strategy was to gather the battle fleet in a large formation, steam forward to a point where the enemy would be compelled to bring out their fleet, use the firepower of dreadnought battleships to overwhelm the enemy battle line, and then blockade the enemy's home waters. This doctrine was the foundation of War Plan ORANGE, the navy's primary strategy for countering Japanese aggression against American territories in the western Pacific. Both war planners in OP-12 and Naval War College students evaluated different paths to reach their enemy, but the basic tenets of the Pacific strategy did not change during the interwar period. What did change was the projected speed of advance. The War College staff nicknamed the predominant doctrine of the early phase as the "steamroller." In later years Edward Miller referred to it as "thruster" or the "through ticket." This doctrine called for the U.S. Navy to steam directly across the central Pacific, engage the Japanese fleet in force, and defeat them as quickly as possible while the garrison defending the Philippines withdrew to the Bataan peninsula and fought a delaying action.[6]

The War College staff established a continuum of relatively unsophisticated strategic wargames that matched the simplistic thruster strategy of the early phase. Wargame records from the first years of this phase, whether ORANGE, BLACK, or RED, bear out the assessments of Spector and Vlahos by describing one-dimensional affairs that provided a plausible geopolitical back-drop to bring two battle fleets into contact. The opening lecture in the 1923 tactics course presented the traditional Mahanian opinion (which the course instructor characterized as that of "the College") that in a major war, while there might be many "minor actions," there would be only one fleet engagement.[7] Once one of the fleets was defeated, the outcome of the "war" would be a foregone conclusion. Even the large-scale tactical game at the end of the course generally had a single mission that inevitably led to a major fleet action. *Guerre de course* and seizing of land objectives were relegated to the background if they were con-sidered at all, and active participation by other nations, either allied or belligerent, was not simulated.

During the first years of the early phase, the staff and students exercised wartime tactical and technological innovations in the context of Mahanian strategies for sea combat, in what William Pratt biographer Gerald Wheeler described as a "classical tone."[8] Some of the developments experimented with during the early phase were not necessarily new, as they had been exercised or at least introduced in some form during the late war, but their place in naval doctrine became more prominent as the technol-ogy behind them matured. At the start of the phase, the best known of these innovations were naval aviation and submarine warfare. Other less-publicized but equally significant develop-ments in naval doctrine occurred during the years immediately after the end of the war. In 1921 Major Earl Hancock Ellis of the U.S. Marine Corps produced the landmark report *Operation Plan 712—Advanced Base Operations in Micronesia*, which presaged the role the marine corps would assume in amphibious opera-tions.[9] The General Board deliberated plans for expeditionary logistics support that eventually led to the Mobile Base Project during 1923.[10] All of these doctrines, sea-based air power, under-sea warfare, amphibious operations, and expeditionary logistics, were introduced and tested with increasing emphasis during wargames of the early phase.

Another development whose effect was not obvious at the time occurred in 1923 when the navy conducted the first of the annual fleet problems. These represented the live-action counterparts to the simulated conflicts fought on maneuver boards at the War College. The first fleet problems occurred in and around Panama and the Caribbean, and were limited by the small numbers of ships and aircraft available to participate. Fleet planners included roles for aviation in these exercises but only in a notional sense, as there were at this time a limited number of aircraft available in the entire navy.[11] Play of submarines in the fleet problems replicated tactics exercised in the wargames, in that submarines steamed in advance of the battle fleet as part of the screening force. During the early phase, War College staff and students began the process of integrating new tactics and doctrines, as well as lessons learned from fleet problems, into game situations of increasing complexity. Limited numbers of assets were not a problem for the wargames designers, who added, on both sides, ships that did not exist, either to experiment with new types or to simulate situations that were not feasible with the existing force structures.[12] Both the games and the fleet problems had their particular strengths and limitations, but as the interwar period progressed, they came to complement each other as venues for training and testing. Regular interaction between wargames and fleet problems did occur later in the early phase, most notably during OP III-27 when the War College replicated the final phase of Fleet Problem VII in the maneuver rooms in Luce Hall. But in the first years of the early phase, such integration of real-world and simulated effort was a bridge too far.[13]

Games conducted during the first years of the early phase reflect a certain amount of "settling down" as the staff reset itself after its wartime experience. The six-month course of instruction gave way to a yearlong course starting in 1921. The game schedule between 1919 and 1922 lists a series of strategic and tactical problems that the staff repeated, with some modifications, from year to year. The percentage of BLUE versus RED games during these years was high, comprising almost half the problems scheduled and played. In 1922 the Strategy and Tactics Departments instituted a more rigorous continuum of games with a higher percentage of ORANGE scenarios, which was fully in place for the War College class of 1923. This class was something of a watershed for several reasons. It was Admiral Sims's final class before his retirement. It was the largest War College class up to that time, and it was the last class before the junior course for lieutenants and lieutenant commander line officers was instituted. It counted as members three future naval leaders of the Second World War: Commander Chester Nimitz, Captain Thomas Hart, and Commander Harold Stark, who was chief of naval operations from 1939 to 1942. Commander Roscoe Mac-Fall, who Nimitz later credited with first devising the circular formation, was also a member.[14] On the staff side, as mentioned in chapter 1, Captain Harris Laning was in his first year as the Tactics Department head for this class. His opposite number in the Strategy Department, Captain Reginald Belknap, was famous throughout the navy as the commander of the task force that laid the North Sea mine barrage during the war. In the staff and class picture from 1923 (figure 14), Nimitz (54) is at back row center and Stark (38) and MacFall (39) are in the third row, sixth and seventh from the left, respectively. The senior students and staff are in the front row, with Thomas Hart (15) second from right, Laning (7) seventh from left, War College president Admiral Sims (9) eighth from right, and Belknap (10) to the immediate right of Sims.

In addition to capturing future naval leaders early in their careers, figure 14 also displays the demographics of a typical early phase class. Scattered among the navy officers are four army, four marine corps, and two coast guard officers.

The games played by this class reflect the state of naval thinking in the early phase and serve as a benchmark for how profound a transformation was to come in the later phases. Played in the second half of September 1922, Tactical Problem II (Tac. 92) was a simple RED versus BLUE game with a full complement of battleships on each side but no submarines or aircraft. The BLUE mission was simply to bring RED to a decisive engagement. Both sides were evenly matched and adhered to the same strategic and tactical doctrines (including traditional linear battle force formations), so the RED and BLUE designations could

FIG. 14. A photograph of the Naval War College staff and class of 1923.
Courtesy of the Naval War College Museum Collection.

have easily been reversed.[15] Strategic Problem III (Strat. 74), an
ORANGE versus BLUE game, introduced a detailed geopolitical
backdrop including the stances of other nations. None of these
other nations figured actively in the maneuver itself, which simulated an American defense of the Philippines by the small Asiatic
Fleet against a large Japanese expeditionary force. Students playing the ORANGE role in Strategic Problem III conducted one of
the first, if not *the* first, experiments with a circular formation.[16]

The development of the circular formation is well documented
in various Nimitz biographies, but it is a useful example of how
even junior students could introduce tactical innovations in the
wargaming venue. As Nimitz biographer E. B. Potter described,
"In setting up the board one day, MacFall placed the supporting
cruisers and destroyers in concentric circles around the battleships. The obvious advantages to this arrangement were that it
concentrated antiaircraft fire and that the direction of the whole
formation could be changed by a simple turn signal, all ships turning together."[17] Potter's interpretation of MacFall's motivation is
less important to this book than the fact that the formation was
MacFall's idea and that he, despite being one of the junior members of his class, was able to suggest it and have it tested on the
War College maneuver board.

Minor games soon gave way to much more complicated maneu-

The Early Phase

vers. The class of 1923 played Tactical IV / TAC 94 in the fall of 1922. Tactical IV, also known as the "Battle of Emerald Bank" after a seamount off Nova Scotia, was the third board maneuver in the class program. It pitted the full-up BLUE fleet against RED, the strongest fleet in the world at the time. As with STRAT III-23, the TAC-94 scenario was only window dressing to the fleet engagement. It started with the BLUE fleet gathered in Narragansett Bay while their RED opponents formed up in Halifax harbor.[18] A RED expeditionary force was "offstage right": in the script but not in the battle. There were no significant landmasses to obstruct the fleet maneuvers, and no land objectives in the battle at all. At the commencement of moves, these two fleets sortied out into the open ocean in large but precise formations just as Harris Laning prescribed in "The Naval Battle."[19] Their respective missions were simple—to find their opponent and defeat them in a classic battle line action. The opposing fleets each possessed subtle advantages over the other. BLUE possessed superior numbers of light forces (cruisers and destroyers), while RED had superiority both in battleship numbers (16:18) and total firepower.[20] Both BLUE and RED had aviation-capable ships (five and four, respectively) in their fleets as allowed under the Washington Treaty, though there were no aviation-experienced students in the class.[21] Submarines on both sides preceded their respective fleets in protective screens instead of maneuvering and operating independently.[22] The scope of the game was Mahan in its purest form—any other tactical or strategic considerations besides finding the enemy fleet and engaging it were ancillary. In a sense Emerald Bank was the Jutland that the U.S. Navy had never experienced in real life.

Professional draftsmen on the War College staff produced blueprint records such as figure 15 to document the sequence of game moves. These charts were essentially time-lapse snapshots of the maneuvers from the "God's eye" perspective and corresponded to the scale-model game board shown in figure 6. A ship's initial and final positions during a given time were plotted as dots connected with a line that represented the ship's track. A small arrow on the line indicated the direction of movement. The chart is not a complete representation of the battle, and it omits much information about the fighting condition of each ship. On the other hand, it does capture a series of snapshots of the overall situation and, in both form and function, accurately represents what a fleet commander would see plotted on a chart aboard his flagship. While the War College students were not yet fleet commanders, they were all used to seeing battle and maneuver information displayed in this manner. These charts were salient features of each maneuver record and served as focal points for postmaneuver critiques.

The record of moves for the approach phase of Tactical IV described two fleets cautiously feeling their way toward an engagement. Each fleet commander arrayed his ships in concentric circles, with the idea that such formations increased tactical flexibility over long columns such as those used by the British and Germans at Jutland. Figure 15 shows move 1, at game time between five and six o'clock in the morning. The RED fleet is in the upper right on a southerly course. The BLUE fleet is at lower left, changing course to the northeast. Note the parade ground formations with battleships in the center and the presence of submarine screens (labeled SS) in geometrically precise rows ahead of each formation.

A curiously anachronistic aspect of the game at this stage was that both teams commenced active maneuvering to obtain the "weather gage," which is a sail-age fighting term for having the wind behind one's own force. Possession of the weather gage was vital for sailing ships, as it provided much more freedom of movement. Early in the steam era, naval officers still considered it an important advantage, as following wind could supposedly extend firing ranges. In their efforts to gain this advantage, both TAC 94 fleet commanders seemed to have been afflicted with a sort of tactical myopia. Using their submarines and aircraft for scouting, both sides located each other at about the same time early in the problem.[23] But once the respective student commanders received the position reports, both RED and BLUE players overlooked or ignored the offensive capabilities of their nonsurface assets for the rest of the problem. The two formations thrusted

The Early Phase

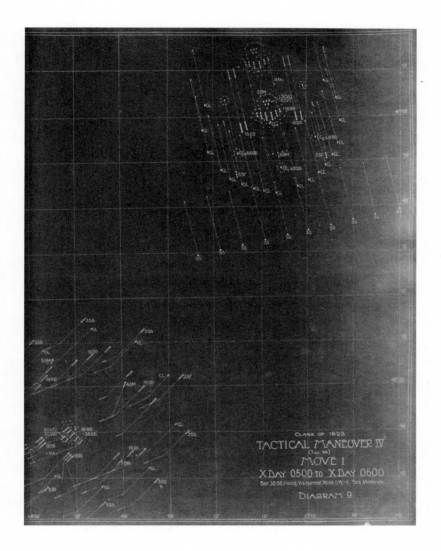

FIG. 15. The first move of Tactical Maneuver IV for the class of 1923. This diagram shows the RED fleet in the upper right, in a circular formation on a steady course of south-southeast. The BLUE fleet is in the lower left and has just turned to the northeast. U.S. Naval War College "Battle of Emerald Bank, Tactical Problem IV, TAC 94, 1923, Diagram 9," Folder 832, Box 19, RG 4, Publications, 1915–77, NHC.

and parried inconclusively with their light forces, and while the BLUE battle line was ultimately successful in gaining the weather gage, by that time the wind was down to four knots and the benefits were negligible.[24] The BLUE battleships eventually came within range of the more powerful and numerous RED battle line during moves 63–65, and at that point the game turned into a shootout whose outcome was a foregone conclusion.[25]

The results of TAC 94 were an unpleasant surprise, at least to players on the BLUE side. RED outnumbered BLUE in battleships by only a small margin but inflicted a crushing defeat. At the end of the maneuver, the staff umpires ruled that all BLUE battleships were sunk or disabled.[26] With the benefit of hindsight, the BLUE team could have prevailed at Emerald Bank if their side had surprised RED with their light forces, but as soon as each force located the other, the outcome became a matter of gun range and weight, with RED holding all the advantages. Thomas Hart's reaction to this game is of interest. While his journal entries when the maneuver started at the beginning of December show that he had high hopes for his side and was pleased with his early performance as a BLUE submarine element commander, by the end of the game in mid-January, he was tired and discouraged. He wrote, "We finished up a very tedious war game today—had been at it for weeks. A fight between our Navy and the British. The latter won, as it usually does in these games and we conclude that we did not get equality as [a] result of that 'limitation treaty' of 1922."[27] Hart's last sentence in the diary entry can be interpreted in two ways. Either he is being ironic about his class's reaction to the game, or he could be genuinely critical of what he considered inequities in the Washington Treaty. In any case, his comment should not be dismissed as mere sour grapes. It indicates that Hart and at least some of his classmates believed the game accurately replicated real U.S. and British force ratios, as opposed to something artificially injected by the scenario developers.

Laning's history and tactical critique of the Emerald Bank maneuver was direct and incisive. While he had no more aviation experience than his students did, Laning at least understood that the ability to locate one's enemy at long range—and to deny that advantage to your opponent by driving off his scouts—would provide a major advantage in battle. He took both sides to task for not taking better advantage of the scouting capabilities of their submarines and aircraft, and he judged the aircraft of both fleets to have been operated "in a rather haphazard manner."[28] The history of maneuver bears his opinion out. Table 2 represents an early attempt to use game results data as part of a wargame critique to emphasize a particular point. It lists RED's inventory of aircraft at the end of the maneuver, recording that almost half of these aircraft ended up in the water for lack of fuel. This result indicates that while the student role-players knew that aircraft could be *launched* at sea, they had not yet considered how to *recover* them in the midst of a battle.[29]

Table 2. TAC IV/TAC 94 RED air order of battle, game conclusion

Status	Torpedo bombers	Fighters	Scouts	Total
En route to Halifax	1	0	3	4
Airborne but picked up later	0	3	25	28
Ditched but being picked up	5	35	8	48
Aboard carriers—not launched	0	21	0	21

A table showing the air order of battle for the RED fleet at the conclusion of the TAC IV/TAC 94 (Battle of Emerald Bank) maneuver for the class of 1923. Note the number of aircraft ditched in the water. This table was created from the data shown in charts that were part of the original maneuver results. *Source*: U.S. Naval War College, "The Battle of Emerald Bank as Maneuvered at the U.S. Naval War College by the Class of 1923: History and Tactical Critique," 80.

The final maneuver for the class of 1923 was Tactical V / TAC 96, also dubbed "the Battle of the Marianas." While this BLUE-versus-ORANGE scenario was more detailed than the Emerald Bank game, it was still a basic Mahanian through-ticket situation played out in a series of maneuvers over a longer stretch of space and time. The BLUE mission for Tactical V was completely in accordance with War Plan ORANGE of the era and exercised a dash directly across the central Pacific through the

Japanese Mandate islands.[30] Arrangements and assignment for this game began in January 1923, and the class of 1923 played it in stages through March. Like other final games, it began as a chart maneuver, evolved into a strategic game (S.76 STRAT V), and ended up on the game floor as a tactical game. As in previous games, students played the ORANGE force role.[31]

While the Emerald Bank maneuver could possibly have been won by BLUE, the situation in the Marianas maneuver was definitely stacked in favor of ORANGE. BLUE not only had to run a gauntlet of ORANGE bases in the multiple island chains between Hawaii and the Philippines but also had to do so while escorting a ponderous train of 170 ten-knot supply ships carrying fuel, ammunition, and other stores. In his maneuver critique, Harris Laning noted that all the advantages in this maneuver—speed, defensive firepower, geography—lay with ORANGE, with one exception. Laning judged that in the open ocean, BLUE had the advantage in air power, but as was previously demonstrated in the Emerald Bank maneuver, this was the weapon that students in the class of 1923 understood the least.[32]

The climax of Tactical V was a night engagement between the light forces of both sides, which started when ORANGE destroyers attempted to break through the screen around the BLUE expeditionary train with a torpedo attack. The attack cost ORANGE twelve of seventeen ships, but their thrust not only reached the BLUE train but also threw the BLUE formation into confusion. Laning's critique, with its descriptions of "desperate" maneuvers, near misses, collisions and near collisions, and "tremendous" exchanges of gunfire reads very much like descriptions of the First Naval Battle of Guadalcanal nineteen years later.[33] The situation at game's end was one that would become commonplace for BLUE-ORANGE games of this period. Both sides engaged in prolonged attacks and suffered substantial losses, but BLUE ended up at the disadvantage. They were battered, a long way from resupply and repair, and still shepherding a large and vulnerable expeditionary train whose objective was still well over the western horizon. In his diary Thomas Hart recorded that the "sons of heaven" (meaning ORANGE, his team) "were in the lead" at the conclusion of the Battle of the Marianas.[34] The staff's maneuver critique was less decisive in assigning a victor, but an important point here is that in both the Emerald Bank and the Marianas wargames, the result was something much less than a clear-cut victory for the BLUE side. The scenarios in the maneuvers may have been unrealistic in hindsight, but they were not contrived to produce a result in BLUE's favor.

Through the interwar period, the head of the Tactics or Operations Department generally wrote the game critique and lessons learned. The author of the Tactical V critique—in this case Harris Laning—developed the "lessons" of the game himself. Input from other members of the staff might have been incorporated, but student explanations or comments are not included.[35] This is not necessarily an indication that there were no comments, and it should be emphasized here that staff and students were frequently equals in rank, especially in the senior course. Later in his time at the War College, Laning upgraded game procedures and record keeping, but in 1923 he was in his first year as head of the Tactics Department and was only one year removed from being a student himself.[36] His postgame critique for Tactical V was rigorous, incisive, and frank, but from a modern perspective, it was also completely "in the box." With the curriculum so structured toward Mahanian doctrine and the estimate of the situation process, it is understandable that so much of his critique focused on adherence to class-taught doctrine and rules. Laning evaluated the estimates of the situation for each team's commander in detail, and gave these estimates a weight equal to that given to the commander's actions. He placed considerable emphasis on a fleet's ability to stay on the offensive, though he also judged BLUE's defensive posture at the end of the game as "sound."[37] While the two documents were of approximately the same length, Laning's critique of the Marianas battle is somewhat spare in comparison with the Emerald Bank critique. It does highlight, however, the most important learning points of the maneuver and is one reflection of Laning's maturation as an instructor.

As the early phase continued, the wargame schedule at the

War College evolved into a generally standard progression similar to the one shown in table 1 of chapter 2. The staff presented the same general sequence of RED and ORANGE games, but the maneuver problems became more complicated and expansive. The 1923 Battle of Emerald Bank evolved into the Battle of Sable Island, which remained a feature of the wargame schedule for the rest of the interwar period. Sable Island was still a Jutland-like fleet engagement in the North Atlantic, but it featured a progressively greater role for naval aviation. The class of 1925 played Tactical Problem II / TAC 98, a new RED scenario situated in the Caribbean. The classes of 1925 and 1926 also played some situations beyond pure fleet engagements, such as defense against landings and convoy protection problems. The maneuver naming convention changed as well, with the former strategic and tactical problems combined under the new heading of "Operations," abbreviated as OP. The trends in wargaming were not completely progressive; the dearth of postgame documents from 1924 to 1927 in the War College archives and the reduced level of detail in those that do exist indicate that Harris Laning's successors in the Tactics Department were less thorough in their capturing of game critiques and lessons learned.

The most ambitious War College exercise of the early phase occurred in 1927, when the real and imaginary BLUE fleets came together during simultaneous, coordinated play of a wargame and the annual fleet problem. This final class of the early phase was, like the class of 1923, one that brought together influential naval doctrinaires on the staff and future World War II leaders among the students. Rear Admiral William Pratt, discussed in chapter 1, was in his first full year as president and reorganized the faculty into a divisional structure that mirrored operational fleet staffs. Future War College president Edward Kalbfus was one of the senior navy captains among the student body. Two of his classmates were Raymond Spruance and wartime Atlantic fleet commander Royal Ingersoll. The junior class of lieutenant commanders and below included naval aviator Lieutenant Forrest Sherman. Figure 16 is the 1927 class picture, showing Pratt at center; Kalbfus in the second row, fifth from left; Spruance

FIG. 16. A montage of the staff, faculty, and students of both the senior and junior classes of the Naval War College class of 1927. Courtesy of the Naval History and Heritage Center.

in the third row, far left; Ingersoll in the fifth row, second from right; and Sherman in the bottom row, third from right.

Like the class picture from 1923, figure 16 illustrates student demographics of the time. The forty-four officers in the senior class included three representatives each from the army and the marines, while the twenty-six-member junior class included two marine corps officers. Sherman was the only aviator in either class.

The class progressed through the usual schedule of lectures, demonstration games, and tactical problems until 2 April, when the staff issued general and special situation descriptions for OP III. OP III-27 was significant in that it not only was coordinated with a live exercise but also was focused on land objectives. The live exercise, Fleet Problem VII, was conducted in stages. It started as a mock defense of the Panama Canal and eventually worked its way up the East Coast of the United States past New York. Amid considerable public interest, this final stage occurred

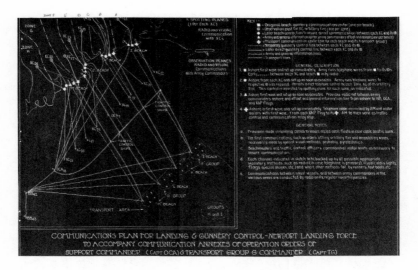

FIG. 17. A blueprint of the communications plan for landing and gunnery control details for the BLACK force during the OP III 27 maneuver. This diagram is indicative of the increasing levels of sophistication achieved in maneuvers of the last years of the early phase. U.S. Naval War College, "Operations Problem III-1928; Summary of Presentation of Staff Solution of the Communications Features of This Problem with Diagrams," Diagram B, Folder 1381-P, Box 39, RG 4, Publications, 1915-77, NHC.

in Narragansett Bay 17–20 May as part of the joint army-navy minor maneuvers.[38] The scenario for these maneuvers called for the BLUE fleet to assist the BLUE army in repelling a BLACK assault on the Rhode Island coast.[39] While the ships were marshaling off the coast, their movements were also plotted in the maneuver rooms at Luce Hall. The unknown writer of the problem's situation description described the Mahanian decisive sea engagement as having already occurred, and the amphibious operation moved from its former place offstage to center stage. The complicated and detailed situation description was much more devoted to the land order of battle than to the maritime, which suggests that army and marine corps faculty and students had much more prominent planning roles than in previous games.[40] Figure 17 illustrates part of the communications plan, which reflects the detail to which the college developed the BLACK landing scheme.

The Early Phase

Positions of forward observers, communications centers on the beaches, and telegraph cables connecting units ashore with fire support ships, backup signal schemes, and spotter aircraft are plotted for all landing sectors. Other equally detailed charts documented the embarkation and debarkation plans, the cruising disposition, the fleet disposition, and the army equipment (including the number of pack animals) carried. When this final phase of the fleet problem ended, fleet commander in chief Admiral Charles Hughes conducted the exercise debriefing at the War College on 24 May, just three days before the class of 1927 graduated. This debriefing marked the first time that a critique for a major fleet exercise had been hosted by the War College. The game itself was successful enough to be repeated for the class of 1928 as their OP III.

The story of the games of the early phase is all about growth. If the games as a venue for experimentation and war planning had not yet reached full maturity, they had at least passed from infancy in 1919 to a sort of adolescence by 1927. At the start of the phase, the movements of the fleets in the games reflected a combination of traditional Mahanian doctrine and lessons that students gleaned from the pages of *The Naval Battle*. Teams limited their objectives to the defeat of the opposing fleet. Each fleet maneuvered in formation, and writers of the critiques paid considerable attention to basic tenets such as concentration of force and archaic factors like the weather gage. Historians who describe the War College wargames as irrelevant rituals are correct if they look no farther than the first years of the early phase.

However, these years were not completely bereft of development. There were still prescient tactical lessons learned in the areas of logistics and aviation. In TAC V 23 the BLUE force experienced considerable problems in shepherding their ponderous fleet train, and Harris Laning's critique identified the speed of the train the limiting factor in the fleet's advance.[41] His recommendations regarding the train were restricted to decreasing its size and increasing its speed, but the growing emphasis on logistics in the War College games and later in the curriculum can be considered progressive by itself. No previous campaigns in American military history featured such long and opposed transits, as well as the need to bring such a high volume of supplies forward to match the battle fleet advance as called for in War Plan ORANGE. For naval officers who had never had to deal with logistics matters beyond their own ships' lifelines before, having to plan for keeping a fighting fleet—even one consisting only of model ships—supplied across a hostile ocean must have been both a new and somewhat daunting experience. The results of their efforts eventually contributed to the later development of fast oilers and ammunition ships designed to steam with the fleet and resupply it at sea as opposed to in port. In 1927 these developments were a long way off, but the interwar War College games and the concurrent fleet problems provided a venue for students to experiment with a variety of different methods of dealing with this challenge,

Experimentation in the field of naval aviation was similarly tentative through the early phase, but this situation should be considered in the context of the technology available at the time. The capabilities of aircraft during these years were minimal, and the whole concept of carrier-based aviation was still truly an experiment. Aircraft bomb loads were too small for them to play a credible role in an offensive strike, so aircraft played primarily a scouting role for the battle line. The potential of aviation was not discounted, however, by all of the class and staff members. Harris Laning may have been a traditional battleship sailor, but his comments in the TAC V 23 critique show that he was at least ahead of his students in his grasp of the value of air superiority.[42] Awareness of the capabilities of naval aviation increased throughout the early phase, culminating in the expanded role assigned to a fully capable USS *Langley* in Fleet Problem VII and OP III 27.

In contrast to aviation, undersea warfare technology figured prominently from the first early phase games, but the development of submarine strategy and tactics did not match that of aviation. In fact, faculty and students probably *overstated* submarine capabilities in these games, which is surprising given their own real-world experiences. In accordance with staff and naval doctrine direction, early phase War College students integrated submarines with the battle fleet in a screening role for which they were ill suited. Submarines did not conduct independent commerce raiding in the wargames as they did during the late war or would do in the war to come, and merchant ships of hostile or neutral nations did not appear in these games except as part of the fleet train. Laning's critique for the Battle of the Marianas emphasized the ORANGE advantage in submarine forces at the close of the game, down to the number of torpedoes remaining, but even this section of the critique shows more emphasis on things that the staff could count than on results.[43]

In summary, War College records show that the first eight years after the end of World War I were development years for wargaming. The continuum of games was established, and while student-officers exercised wartime innovations in the context of Mahanian strategies during the early phase, the stage was set for experimentation beyond those boundaries and expansion of experiments in strategy, tactics, and technology. The Naval War College and its wargames passed from infancy to adolescence, and both were about to enter a phase in which their potential as venues for experimentation would be much more fully realized even as the navy itself endured a period of sharp decline.

4

The Middle Phase, 1928–34

The first nine years of the interwar period at the Naval War College were a time of growth and development. The "middle phase" years of 1928 through 1934 were a time of increasing relevance for the War College, as the school continued to put aside nineteenth-century notions of naval warfare and expanded the testing of new strategies, tactics, and technologies in a maturing environment of wargaming. The college staff, faculty, and students were building their wargames into increasingly useful engines of preparation for war during seven years of precipitous decline for their service.

In contrast, if any period of modern history can be considered the nadir of U.S. Navy fortunes, the middle phase has to be among the leading contenders for the position. All branches of the American military felt the effects of the Great Depression, but the navy—being particularly tied to the American industrial base—reflected the impact in both numbers of active ships and new shipbuilding. Because of reduced military expenditures and then the economic downturn, the American naval shipbuilding industrial base became largely inactive. Shipyards in Brooklyn, Philadelphia, and Newport News went from 1919 until 1927 without laying one cruiser or destroyer keel.[1] Overall, naval strength measured as the total number of surface combatants (battleships, cruisers, and destroyers) declined from a 1919 high of 230 to an all-time low of 119 by 1931. The Washington Treaty of 1922 allowed the conversion of two battle cruisers already under construction to aircraft carriers, but limited budgets for shipbuilding meant that there were only three carriers in the U.S. Navy by the time of the stock market crash in October 1929. At the end of 1933 the navy consisted of 372 ships of all types displacing 1,038,660 tons, which was 150,000 tons short of the Washington Treaty limitation. To man these ships, the navy could muster 5,929 officers and 79,700 enlisted, which restricted the manning of even this reduced number of ships to only 80 percent strength.[2] By failing to build up even to the treaty limits, the United States had fallen far behind the other major naval powers in new ship construction. Figure 18 illustrates this disparity in both ship tonnage and the number of ships appropriated by the Washington Treaty signatories between 1922 and 1933. If one accepts the well-known characterization by Japanese admiral Isoroku Yamamoto of the United States as a "sleeping giant" at the start of World War II, then these U.S ship appropriation figures are a good indication of the depth of that sleep. Yamamoto referred to the U.S. industrial base, and the naval shipbuilding segment of that base did fall into a prolonged slumber that would take the U.S. Navy years to overcome.[3]

Concurrent stagnation of shore base construction and support had a particularly significant impact overseas. As part of the effort to persuade Japan to accept a Washington Treaty tonnage restriction that was proportionally less than that of the United States or Britain, the U.S. delegation agreed not to militarily strengthen any Pacific bases west of Hawaii. This arrangement was crucial to the outcome of the disarmament conference, but it halted base improvement initiatives in the Philippines and on the islands of Guam, Wake, and Midway. Like the dormant shipbuilding program, lack of modern defenses on these territories would have a major impact on American ability to defend them in 1941.

In some ways, such as its aging physical plant, the Naval War College reflected the declining material state of the navy. Conversely, the records of wargames played during the middle phase years show that the college remained a vibrant and influential center of strategic and tactical thought, and even *increased* its influence, within the navy. Enrollment in the senior course remained at a steady level during the middle phase years, and

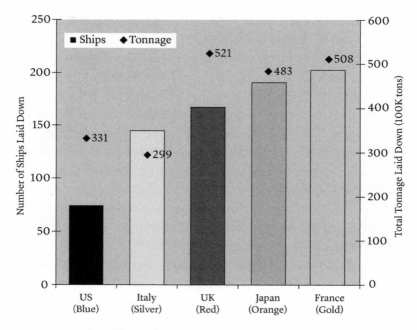

FIG. 18. A chart illustrating the number of ships and the total tonnage appropriated by the Washington Treaty Powers during the years from 1922 to 1933. Of all the treaty signatories, the United States is the lowest in terms of the number of ships appropriated and the second lowest in terms of tonnage appropriated. *Annual Report of the Secretary of the Navy for the Fiscal Year 1933* (Washington DC: Department of the Navy, 1933).

the size of the junior course expanded from an early phase average of twenty-three to thirty-nine. Together with the institution of the correspondence courses in 1930, this growth spread War College lessons through a wider percentage of the officer corps. The college faculty retained the reputation they established in the early phase by continuing to attract a number of officers who would eventually achieve flag rank. After Admiral Sims's retirement in 1923, the navy established the term of the War College president as three years, and a succession of alumni and former faculty served in this capacity. Foremost among them was Admiral William Veazie Pratt, president from 1925 until 1927. A number of notable naval officers took seats on the faculty during the middle phase years. Captain Joseph Taussig returned to the staff as Pratt's chief of staff and war plans division head. Navi-

The Middle Phase

gation expert Captain Benjamin Dutton '29 moved from the student to staff ranks after his graduation. Commander Raymond Spruance, a 1927 graduate, joined the faculty as director of the correspondence course in 1932.[4]

Admiral Pratt was arguably the most proactive of the interwar period college presidents, and the college took major steps toward the naval mainstream during the years he held the office. With a view toward his students' future assignments, he restructured the school organization at the end of the early phase into divisions that mirrored the operational fleet and Chief of Naval Operations staffs. He introduced the "committee" approach to problem solving, stressed the study of logistics, and revised the curriculum to remove the split between strategy and tactics. Pratt's new Operations Department dispensed with the early phase wargame designations of Strategic and Tactical in favor of the more generic Operational maneuver. The east and west game rooms in Luce Hall remained the arenas for the board maneuvers, but faculty game critiques reflected a growing awareness that the physical limits of the rooms limited the scale of games they could accommodate.[5]

More significant than organizational changes were the ways that the games broadened in scope. Several new strategic and tactical innovations in the greater navy—some in development and others that were only conceptual—found their way into the games. The game descriptions and critiques written by members of the Operations Department during the middle phase reveal increasingly sophisticated scenarios that included aspects of naval warfare beyond battle line engagements, such as amphibious landings and raiding. Improvements that staff members instituted in game procedures increased game speed, which in turn allowed for longer game periods, broader geographic settings, and a wider range of both objectives and the approaches used by students to achieve those objectives.[6] The new quick-decision maneuvers, already mentioned in chapter 2, were first assigned to the class of 1929.

Experiments in strategy, tactics, and technology expanded during the middle phase. These virtual sea trials reflected a continuing evolution of both the vision of a Pacific war and the wargame processes for simulating that conflict. On the strategic level, the thruster strategy of the 1920s was challenged by a new "cautionary" (another Edward Miller term) strategy, which assumed that American territories in the Far East would fall to an initial Japanese offensive and would be regained only after a protracted, systematic advance across the Pacific. The increasing imbalance of U.S. and Japanese naval force structures, results of fleet problems at sea, and a growing body of evidence from previous wargames at the War College reinforced this pessimistic outlook. Accordingly, the Operations Department instructors wrote strategic and tactical scenarios for BLUE versus ORANGE games that began with an assumption that BLUE had already suffered a major setback or that the Philippines had already been lost and the BLUE mission was a recapture of lost territory, as opposed to a relief of the Philippine garrison.[7] Some maneuvers were continued from one class to the next, reflecting the prospect of a longer, harder battle to cross the Pacific. With longer scenarios, games took on a different character, with emphasis on achieving long-term objectives rather than winning one battle.

The War College staff did maintain the original Philippine focus of the ORANGE games. But while the objective of the BLUE fleet was to establish advance bases at either the island of Tawi Tawi or Dumanquilas Bay, they experimented with routes to those positions from all points of the compass. Approaches from the north, along the Aleutian Chain and ending in Avacha Bay on the southeastern coast of Kamchatka Peninsula, were tried as early as 1922 and continued as late as 1934.[8] More common approaches included the Great Circle route north of the Japanese Mandate islands (the shortest route) or straight through the Mandates via the island of Truk (the most supportable route). By the end of the middle phase, BLUE players departed completely from War Plan ORANGE convention by sailing their imaginary fleet *east* across the Atlantic, through the Suez Canal, and across the Indian Ocean to the Java Sea.

Tactically, single decisive engagements such as the 1923 Battles of Emerald Bank and the Marianas gave way to campaigns with several engagements, some very costly but few that were really decisive. In 1986 Michael Vlahos interpreted this shift as reflecting a growing appreciation of "strategic geography" as opposed to focusing on ship tonnage ratios, with the attendant realization that attrition would replace the decisive action.[9] Movement records and game critiques started to reflect increasing awareness of and emphasis on the potential of carrier-based aviation. Instructors began to admonish students to avoid "geometrical movements" such as those performed by the class of 1923.[10] Refueling at sea became a standard tactic, though the lack of fast oilers in the real navy meant that the large battleships refueled smaller ships during maneuvers on the game board.[11] Amphibious landings to secure island bases evolved from minor expeditions to major operations. Submarines still coordinated their operations with the surface fleet, but student players on both sides detached their submarines from the fleet formation and had them range farther afield. Games lasted longer and ended when the clock ran out, not necessarily when one fleet was vanquished.

Technically, naval assets that were neither built nor even designed when the games were played found their way into fleet lineups. Staff officers charged with writing the maneuver scenarios added rigid airships (ZRs) to the BLUE force lineup to scout ahead of the main body in OP IV-28, three years before *Los Angeles* (ZR-3) participated in Fleet Problem XII.[12] They considered floating dry docks to accommodate repairs at advanced bases in OP VI-29 when the Mobile Base Project was still a paper concept and four years before the first auxiliary repair dry dock was launched in 1933.[13] Recognizing that large-scale wargames required more air power than the three U.S. carriers of the day could provide, the Operations Division first added auxiliary aircraft carriers converted from passenger liners, and later a hybrid cruiser-carrier designated CLV to the BLUE force mix. The CLV first appeared in OP II-31 SR.[14] Its design (figure 19) provides a window into the mind of interwar naval planners. First, it circumvented the Washington Treaty restrictions in that it fell under the unrestricted light cruiser category and not the carrier category. Second, it featured substantial armament in three gun

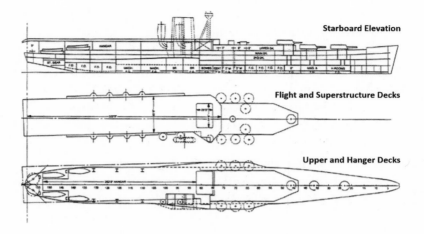

FIG. 19. A 1931 plan for the Flying-Deck Cruiser (CLV) concept, showing the starboard elevation, the flight deck, and the hangar deck. CLVs were intended to carry a small number of aircraft but operate as light cruisers after their aircraft were launched. Their classification as cruisers meant that they fell outside the terms of the Washington Naval Treaty. Andrade, "Ship That Never Was," 136. The CLV plan is from General Board Study #1516, 420-8 (cruisers), 26 January 1931. Three later designs for a "Proposed Flight Deck Cruiser" appear in "Bureau of Ships Spring Styles Book #3 (1939–1944)," Record Group 19, Records of the Bureau of Ships, 1940–66, ARC Identifier 1696046 / Local Identifier 019SPRINGSTYLES3&4, National Archives and Records Administration, Washington DC.

turrets forward, and an angled flight deck capable of operating some eighteen aircraft extending from amidships to the fantail.

The CLV deck was angled to port but not to allow simultaneous launch and recovery operations as are modern aircraft carriers. The angle would have served to keep launching aircraft clear of the gun directors and other equipment in the forward part of the ship.

The cruiser-carrier was the subject of navy testimony in front of the House Naval Committee and remained a topic of considerable discussion in naval circles until 1940, though it was never actually built.[15] Nonetheless, it was a modern design of a multirole ship, and if one substitutes missile launchers for the gun turrets, the CLV design is strikingly similar to the Soviet *Kiev*- and British *Illustrious*-class cruisers of the Cold War.

The Middle Phase

The first and next-to-last of the classes of the middle phase provide good examples of how the games evolved as laboratories for experimentation as the interwar period continued. The outline of course for the class of 1928 stated that the year's work would investigate whether a rapid thruster-like advance across the Pacific could be supported logistically, or if such an advance should be delayed until the necessary supply, maintenance, and repair capabilities were in place.[16] The schedule for the class provided for a logical progression of maneuvers that introduced students to situations of increasing complexity. OP I-28 was the defense of a fixed point (a base) against an attack. OP II-28 was the defense of a moving point (a convoy) against attack. OP III-28, the conduct of an opposed amphibious landing,[17] is a significant game for two reasons. First, it was a BLUE versus BLACK game connected with the final stages of Fleet Problem VII, but the description of BLACK sounds very much like RED. Second, BLACK is the attacking force, and the centerpiece of the maneuver is the complexities (the "special situation" in the game description) of the BLACK problem, not BLUE's defense.[18] In these introductory games the color designation of the opposing sides and the specific geography were less important than the overall situation students found themselves in and the mission they were assigned to accomplish.

OPs I, II, and III led to OP IV-28, which was a long, complex game that combined specific aspects of the three previous situations in the larger context of a war plan. The course outline for the class described OP IV as a problem of the "intermediate phase of the Pacific (Blue-Orange) situation." The author of the game critique considered OP IV to be "the first time in the history of our Maneuver [that] a series of situations was continued in such relation that the Problem was enabled to pass from the Strategical Phase through a natural concentration period into a realistic tactical phase."[19] Student committees studied OP IV-28 from 17 January to 4 February, planned between 6 and 11 February, and maneuvered from 23 February until 21 April.[20] During the maneuver phase, BLUE force players gathered in rooms in the east wing of Luce Hall, while their ORANGE opponents set up shop in the west wing. Every member of the class had an assignment to a committee and to a mock position in the BLUE or ORANGE fleet staffs; captains and senior commanders played the roles of admirals, while the more junior commanders, lieutenant commanders, and lieutenants took positions as leaders of light force (light cruiser and destroyer) divisions. Army and marine corps students held staff assignments that were as appropriate as possible.[21]

From the BLUE perspective OP IV was a classic thruster situation, with the larger BLUE forces advancing immediately across the Pacific before ORANGE could solidify their gains in the Philippines. The BLUE plan was to "advance step by step as rapidly as each step could be made in assured superior strength . . . to establish the Advance Fleet (Battle Fleet reinforced) in TRUK, gain control of the Mandate islands for BLUE use, and operate offensively until the arrival of the Expeditionary Force and Reinforcing Escort."[22] The BLUE commander, Captain George E. Gelm, elected to combine his ships in one large formation, with the train in the center and the fighting forces on the periphery. It was not long before the ORANGE side located this large BLUE force, and the maneuver turned into a protracted maritime brawl. The blueprint record of moves 38 to 40 (figure 20) shows a snapshot of the campaign as it progressed once the BLUE advance and expeditionary forces started their transit from Truk to their ultimate objective, the island of Tawi Tawi, southeast of the Philippines. At this point in the game, the BLUE fleet is proceeding in formation with the expeditionary force ships in the middle, screened on all sides by destroyers. The BLUE battleships are involved in a melee on the southern flank of the formation, where they and their escorts are getting the worst of an ORANGE air and surface torpedo attack. An ORANGE destroyer group has just crossed in front of the BLUE Main Body.[23]

By the end of the maneuver, BLUE had reached their objective of Tawi Tawi and still had ten of their original twelve battleships, but only two of these were undamaged and more than half of the BLUE destroyers were either damaged or sunk. While the smaller ORANGE fleet had suffered similar losses, the college

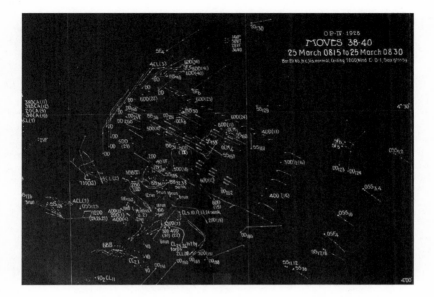

FIG. 20. A blueprint of moves 38 through 40 of OP-IV played by the class of 1928. The BLUE fleet and train are in the center and moving northwest, while the ORANGE fleet is attacking BLUE from the west and south. U.S. Naval War College, "Blueprint Record of Moves, OP IV 1928; Moves 38–40, 25 March 0815 to 25 March 0830," Folder 1382-V2, Box 41, RG4, Publications, 1915-77, NHC.

staff judged that the situation for BLUE was not a win or even a draw but instead was "about as bad . . . as could be expected." The BLUE fleet was battered, had long lines of communication to protect, possessed few undamaged ships, and had no means to conduct even rudimentary repairs.[24]

The lengthy "Conclusions and Lessons Learned" report from OP IV was not signed, but the author was most probably Captain John W. Greenslade '26, the Division C (Movement) chief and the game director of operations. His report differs greatly from Harris Laning's matter-of-fact "History and Tactical Critique" reports for the class of 1923 in its focus on broader, more subjective areas. Some of this difference could be due to improvements in record keeping; the 1928 blueprint records of moves are much more detailed and heavily annotated than the 1923 charts, so it is possible that Greenslade did not feel the need to provide a move-by-move narrative as Laning did. Whatever the reason,

Greenslade's critique is less a narrative than an essay. He started with a rambling recap of the school year and then discussed his philosophy about the relationship between wargaming and war planning, and his view of the benefits of combining strategic and tactical games. He echoed Admiral Pratt by emphasizing the importance of logistics, the specialization of staffs, and the relationships between logistics and fleet commanders, calling the War College emphasis on logistics "prophetic of a change" in the rest of the navy[25]

But once he was finished with this general preamble, Greenslade critiqued the maneuver in specific and blunt terms. He was especially definitive regarding the viability of BLUE strategy. Greenslade declared that while a fleet of sufficient size to defeat ORANGE could be sortied "promptly," the odds of BLUE forces in the Philippines being able to hold out in and around Manila until the fleet arrived were remote unless those forces were greatly reinforced prior to hostilities. Even this possibility he dismissed as "incorrigibly optimistic."[26] Greenslade next went beyond strategy to make remarkably prescient recommendations for changing the tactics and organization of the aviation assets of the relief force to increase their autonomy and overall effectiveness. His comments, coming as they do from a senior officer who spent his career in surface ships, are especially significant, as he emphasized the need for the BLUE surface units to operate under an umbrella of what would later be called air superiority: "It seems established that BLUE can only hope to make successful operations of retaking the PHILIPPINES from TAWI TAWI by wiping out ORANGE air bases by air in succession and pushing forward the zone in which surface craft may operate. . . . All [non BB or CA aircraft] should be organized in an Air Command from which they operate as a Force or with their carriers as parts of Task Forces as designated by High Command."[27]

Greenslade did not restrict his recommendations to the employment of aviation. He also expanded what he considered to be needed improvements in U.S. submarine types: "For the purposes of carrying war at an early date into the Western PACIFIC submarines are the most important type. We cannot have too many submarines of good quality for purposes of attack, screening, scouting, and reconnaissance. . . . Our plans should contemplate maintaining many of them far westward and that they should be sent forward immediately after D day into the western PACIFIC."[28] At this point, it is probably worth recalling that when the class of 1928 played the OP IV game, U.S. Navy real-world experience with aircraft carriers was limited to fleet problems conducted with the experimental *Langley* (CV-1). The newest American carriers, *Lexington* (CV-2) and *Saratoga* (CV-3), were commissioned in late 1927 and were still undergoing shakedowns. Regarding submarine quality and ability to make long patrols, the most numerous American type in service was the one-thousand-ton S-class. In 1921 a group of these "S-Boats" took *seven months* to deploy from the East Coast to Manila.[29] The 1928 navy was not remotely capable of executing Greenslade's vision and would not be until the early 1940s. Greenslade himself had never participated in major operations with aviation and submarines; he based his conclusions on his year as a student and his two years' experience orchestrating RED, BLACK, and ORANGE wargames.

In spite of, or perhaps because of, the grim assessments of the BLUE situation once the Tawi Tawi base was secured, the War College staff attempted to salvage some learning points from OP IV-28. They made the end state of that game the starting point for OP VI-29, a joint effort between the army and naval war colleges. Their objective was "to determine if it would be possible under those conditions to prosecute further naval operations and ensure the safe arrival of additional Army forces and supplies."[30]

The critiques written by John Greenslade make clear that he, like Harris Laning, was a driving force behind the use of wargames as a venue for experiments during his years on the War College staff. The records of games conducted after his return to sea duty in 1928 reflect some loss of momentum in innovation and lack of consistency in gathering and interpreting results. The critique of OP II-31 by Captain Benjamin Dutton '29 is very detailed, complete with blueprints of moves and detailed discussion of fires.[31] But one year later, the critique of TAC I-32 by Captain R. B. Coffey '22 is exactly the opposite. It is high-level, containing florid prose and historical analogies of the Battle of Lake Champlain.[32] Dutton, a career surface sailor, delegated the critique of the air battle to Commander Newton H. White Jr. '30, one of the first qualified pilots to serve on the War College faculty. White performed the same function during the TAC 1-33 critique. This indicates the importance of the faculty members assigned as head of the Tactics and Operations Departments in maintaining and advancing wargames as mechanisms of preparation and innovation.

The analytical character of both the games and the postgame critiques was restored and increased with the return of Harris Laning, who came back to serve as War College president from 19 June 1930 to 18 June 1934. One of Laning's first acts during his tenure was to establish a new Research Department for collecting and analyzing the reams of data generated by the games, and then collating it with similar data from other sources such as the fleet problems. Thomas Withers '24 was the first head of the new department from 1931 to 1932, and was followed by Captain Wilbur Van Auken '27.[33] By the time that Captain Van Auken assumed the lead of the Research Department, the ORANGE trans-Pacific scenario had been played multiple times by each postwar class. While staff and students had experimented with multiple variations in force composition and tactics, the overall results of the games had been generally consistent. Because of the high turnover rate in both students and staff, up to this time game critiques had focused on short-term lessons learned. The Research Department's primary contribution to the college mission was to look at the games in the long term across several classes. With this new branch in place and functioning, the War College became, in Laning's mind at least, "an almost perfect research laboratory for every detail of naval warfare."[34]

Tactics employed in the games of the classes of 1931 and 1932 reflected an increasing awareness on the part of both faculty and students of the potential of sea-based air power as a force multiplier. These games all featured major air strikes launched and absorbed by both sides before the surface fleets ever came into contact. The air commanders on both sides experimented with different priorities for their air attacks.[35] In OP III-31, BLUE targeted RED cruisers while the RED commander elected to target the BLUE carriers. The BLUE strategy misfired. While BLUE air attacks sank five RED cruisers, RED aircraft disabled all three BLUE CLVs and the single CV. In two hours, the air attacks on both sides had spent themselves, and while the BLUE players had a slight advantage at that point, they could not maintain it without air cover.[36] This emphasis on air attacks over surface was such that Admiral Arthur J. Hepburn, the senior member of the senior class of 1931, was compelled to caution his classmates that it appeared that the navy was "heading toward a conception that there will always be an air fight before two surface forces come together, which might be a dangerous conception."[37] But the game records and postgame critiques that survive show that the students continued to expand the use of their nascent air capability well beyond something like manned gun rounds fired downrange and then left to their own devices as they had in 1923, and more like a separate force that was best employed when given less directive guidance.[38] In OP II-32, the BLUE players went so far as to experiment with deploying their carriers independently from the battle force on a scouting line once they arrived in theater.[39] They did not show the same willingness to experiment with their submarines, treating them like "mine fields with some degree of mobility" and minimizing their potential for night operations.[40]

Members of the Operations Department did not neglect the Atlantic theater of operations. Their RED game scenarios also expanded beyond the classic fleet-versus-fleet engagements, though War College classes continued to play the classic Sable Island game until well into the late phase. Letters from Joseph Taussig to John Greenslade reflect that the Operations Department staff spent considerable effort on this expansion, and that Taussig at least saw developments in Europe as important enough to justify a division of U.S. naval power between the Atlantic and Pacific.[41] With these new scenarios, the RED games now allowed War College students to experiment with strategy and tactics in conditions that complimented the ORANGE games. Unlike the

Pacific scenarios, the RED games put BLUE in a defensive role with generally inferior numbers, as the Atlantic-based scouting force was much smaller than the Pacific-based battle force, which contained most of the U.S. Navy's striking power. In OP II-30, the BLUE scouting force faced a larger RED expeditionary force off the New England and Canadian coasts. In OP IV-30, BLUE's objective was to protect approaches to the Panama Canal with the scouting fleet and delay RED's approach through the Caribbean until the battle force was able to conduct their transits. OP III-31 was a similar game, except that RED established an advance base at Trinidad. In OP III-32 the operations staff threw the students a curve by changing political conditions—and the offensive and defensive roles of RED and BLUE—while the game was in midstride.[42]

Another major development in the ORANGE game was strategic in nature as opposed to tactical. Having tried all manner of approaches across the Pacific to the Philippines, the game staff for OP II-32 allowed the students to plan an approach from the opposite direction, across the Atlantic, through the Suez Canal, across the Indian Ocean, and to the Philippines by way of the Dutch East Indies. This far longer route provided a certain measure of safety for the BLUE fleet but also encouraged students to consider the importance of cooperation with allies in Southeast Asia. The Research Department report on this game called the route a "radical departure in the strategic games at the War College."[43]

The class of 1933, like its counterpart ten years earlier, counted some future notables among its members. Foremost among them was Captain Ernest J. King, the senior class member and future wartime chief of Naval Operations, and Captain William F. Halsey, the wartime commander of Third Fleet. Both King and Halsey were latecomers to aviation but had qualified as pilots, instead taking the shorter aviation observer route as Admirals Moffett and Reeves had done years earlier. King's aviation designation and seniority in the class ranks meant that he assumed leading roles in the games, particularly as commander of RED aircraft carriers in TAC IV and as overall BLUE commander in OP IV.

The class of 1933 played the same basic progression of games that their predecessors had experienced. Three of the more significant games played during their class year were OP II-33, TAC IV-33, and OP IV / TAC V-33. OP II-33 was the second experiment in sending the BLUE battle force to the Philippines via the Suez Canal and the Indian Ocean. But while the route taken was different, the result was dishearteningly familiar. BLUE received heavy damage to their battleships and carriers while steaming up from the south, and arrived at Tawi Tawi too late to prevent Luzon from falling. ORANGE was able to attack the BLUE expeditionary force convoy steaming west from Hawaii before it could link up with the battle force. Wilbur Van Auken described BLUE's situations in the Research Department's comments on the game: "In only 45 days BLUE has accomplished her mission of establishing a base in the Western Pacific—but at a terrific sacrifice. ORANGE has been able to work on interior lines, select her operations, receive quick replacements and re-enforcements, been sure of her communications and finds it practicable to effect repairs of damages which she has received."[44]

A major point stressed by the Research Department was that the BLUE fleet's approach route highlighted the previously underappreciated need to receive fuel, provisioning, and basing support from allies in the region, namely England, Australia, and the Netherlands East Indies, and the need to repair underwater battle damage closer to the theater of operations. Beyond tactical concerns, the Research Department also ventured into the diplomatic arena by advocating the reduction or outright abolishment of submarines, and into naval architecture by emphasizing the need for a "sloop of suitable design for anti-submarine duty and patrols which can be produced in large numbers."[45] By 1933 submarines were too far past the point where they could be abolished by diplomacy, but the sloop referred to in the report actually did appear years later in the U.S. Navy in the form of the destroyer-escort (DE). American shipyards eventually produced some four hundred DEs, and these ships filled the role envisioned at the War College nine years before the United States entered the war.

TAC IV-33, a BLUE versus RED game, also presaged certain aspects of the coming war. Captain King played the role of commander of the four RED aircraft carriers in this maneuver, and elected to direct the striking power of his carrier aircraft exclusively against BLUE cruisers. Captain Frank Robert McCrary, the navy's first balloon pilot and King's opposite number on the BLUE side, elected to target RED's carriers. King and McCrary's target selections mirrored those of their predecessors in the OP III-31 maneuver. By the end of the game, King's choice of tactics had cost RED practically all their planes and carriers. BLUE's aircraft losses were heavy, but they kept their carriers intact and retained at least some degree of long-range scouting and strike capability.[46] The class played TAC IV-33 through to a surface engagement, but by this time, initial attacks with aircraft were becoming standard game tactics.

The class of 1933's major game was the BLUE versus ORANGE maneuver OP IV, in which the students planned and executed the approach stages of what R. B. Coffey (now the head of tactics for the senior class) called "our most probable war."[47] By this time, almost every possible route to the Philippines had been evaluated in multiple wargames. One that had not been attempted yet was one that kept well south of the Mandated Islands and skirted the north coast of New Guinea. This route was the staff's preferred solution, and they pressed the BLUE players, led this time by Ernest King, to try it. King considered the southern route to be too exposed to flank attacks, and instead favored the old Great Circle route. In his narrative history of the Naval War College, John Hattendorf described this difference of opinion as being resolved only when King departed Newport early to assume the post of chief of the Bureau of Aeronautics.[48] The game critique describes a less dramatic situation, with King agreeing to play the staff solution of this problem "in order to investigate the possibilities of the route south of the Mandated Islands where ORANGE will also have a logistic problem to consider."[49] Despite the use of the new route, the results of OP IV were the same as in previous games. BLUE was fifteen hundred nautical miles short of their objective in Dumanquilas Bay by the end of the maneuver, and

The Middle Phase

their situation in comparison with ORANGE was precarious. Of the fifteen battleships that BLUE fielded at the start of the game, only seven were undamaged. Two of the four BLUE aircraft carriers were damaged, and half of BLUE's aircraft were lost. The ORANGE battleships had not even entered the fray—all BLUE losses had come from daylight air attacks and nighttime torpedo attacks by cruisers.[50]

The postgame critique for OP IV-33 started with a discussion of specific points of tactics, most notably the relative capabilities of air and surface attacks against capital ships. Army Air Corps major Follett Bradley started with "a brief talk on air fighting. Commented that it was a highly controversial subject." After some polite sparring between Bradley and navy captain Kenneth Whiting about the survivability of bombers in daylight attacks, the critique evolved at the end to a frank discussion of overall strategy.[51] Captain C. R. Train, who played the role of the ORANGE commander in chief, asked bluntly "if it was a good thing for [the navy] to give so much thought to this crossing the PACIFIC when it is pretty well established that it could not be done." Captain Stephen Rowan '21, the chair of the Operations Department and another veteran surface line officer, voiced an opinion that the only way BLUE could successfully cross the Pacific was under the cover of a "great preponderance" of air support.[52] This degree of air capability was something that the real U.S. battle force did not yet possess, but Rowan's comment is another indication of a higher degree of support for naval aviation among senior naval officers of the interwar period than is commonly assumed.

With the same basic results obtained despite experiments with different tactics, new ship types, and a variety of approach routes, commentary by the new Research Department reflected a growing realization that the thruster strategy was too costly for BLUE. At the same time, ORANGE, while in many cases suffering losses that were comparable to BLUE, usually ended the games in a better position to recover from those losses and mount a counterattack. Captain Van Auken's summary of OP IV-33 and TAC IV-33 on this point is worth quoting at length to show his original points of emphasis:

. . . it seems as though the <u>usual</u> heavy losses by BLUE are bound to occur even though ORANGE suffers corresponding ones. . . . Seven years at the War College have brought out the same weaknesses to the front. In each problem, each year, there are "set-ups" to use as examples for BLUE planning, either "a steam roller"– "step by step advance" or "major raids" in a war with ORANGE . . . . Since 1927, with different size fleets, over different routes, with various assumptions, the <u>same</u> points of strategy tactics, types, gunnery, weapons and effect on personnel have arisen. This problem, based on the results of the game and the future ahead of the BLUE fleet in the Western Pacific, again shows the tremendous obstacles to be overcome—even with a Navy constructed up to Treaty Strength.[53]

So, by the end of the middle phase, the War College and the greater navy had reached a strategic crossroads. Repeated experiments with all manner of approaches to the ORANGE war led to the same conclusion—that the U.S. Navy's thruster strategy for the Pacific was not workable with the fleet in hand. The realization summarized in Van Auken's report led the Research Department to conduct a comprehensive review of all of the trans-Pacific games played between 1927 and 1933. The Department compared BLUE and ORANGE losses in each game as a function of the BLUE force's approach route, ORANGE defensive strategy, force composition (either actual or in accordance with the London Treaty), and BLUE force speed of advance. They concluded that variations in each of these factors made little difference in the game outcomes, and that BLUE would be at a major disadvantage in trying to fight their way past ORANGE-held island bases to reach the Philippines, simply because their continued advance put them farther and farther from their support infrastructure.[54] The Research Department sent the results of their review to War College president Admiral Luke McNamee in December of 1933, and the report eventually reached all the way to the desk of the chief of Naval Operations.[55]

Despite the grim conclusions from the Van Auken report, the War College continued to plug away at the trans-Pacific problem.

OP IV-33 was continued in the next class as OP III-34, and the results of that maneuver reinforced the 1933 finding that BLUE suffered its greatest losses from air attacks during the day and submarine attacks at night. Other BLUE vulnerabilities made themselves apparent as well. To add insult to injury, ORANGE succeeded in luring the BLUE fleet away from its advanced base, which they subsequently attacked and seized. This action engendered one particularly prescient comment from the writer of the OP III-34 critique that "a base unable to defend itself is of no strategic value as a base."[56] This game board finding would be proven in real combat at Wake Island and Guam during December 1941.

Despite severe economic pressures that limited the size and capability of the fleet—or perhaps because of those pressures—the games of the middle phase were the venue for some of the most boundary-stretching experiments of the whole interwar period and produced some of the most insightful critiques. When the results and critiques from OP IV-28 are compared with those of OP IV-33 and TAC IV-33, it is evident that the wargames had reached a much higher level of maturity as an experimental venue. The progression of the games within each class and from class to class had become institutionalized, and by 1933, there was a sufficient body of game results to enable the Research Department to develop instantiated observations about not only student performance but also the viability of naval strategies. These observations from mock campaigns and battles in turn affected both strategy and tactics in the greater navy. Some of the more notable experiments, with different ship types, employment of aviation, and use of different routes to approach the Philippines migrated through the students, instructors, and then out to the fleet. This happened in spite of the assertion by the school that the wargames were not laboratories for war plan development. In reality they could not help but be, simply because of their ubiquity and their demonstrated potential to provide a low-cost way to exercise naval thought in an era of parsimony. The game represented one of the only ways that issues of pressing interest to the navy could be investigated in a systematic way, and during the middle phase years, the games truly came into their own. With the start of the late phase, it remained to be seen how the games would help to develop a Pacific strategy that could actually work, instead of continually exercising a strategy that was not viable.

5

The Late Phase, 1935–41

The years from 1935 to 1941 saw a major shift in the navy's place in national policymaking. The rise of expansionist governments in Germany and Italy and their aggressive movements in Spain, Ethiopia, and eventually central Europe drew attention away from the Pacific Ocean to the Atlantic. The Roosevelt administration's efforts to revive the moribund national economy through (among other initiatives) shipbuilding and base construction energized the navy, though the road to combat readiness for a two-ocean war would be a long one.[1] But with this newfound sense of purpose, the wargames at the Naval War College quickly took on an urgency and level of relevance that corresponded to accelerating changes in the war plans. The increasing pace of naval technologies also added more emphasis to the War College role as an experimental laboratory.

The first year of the late phase saw the beginning of Admiral Edward Kalbfus's first term as college president. Kalbfus (class of 1927), a scholarly and cerebral man who remains the only War College president to serve two separate terms, was completely dedicated to the college's mission of teaching decision making, and he was actively opposed to a direct role for the school in war planning. In a speech before the start of the class of 1935's Big Game (OP III), Kalbfus emphasized that the game was not an *official* test of a war plan but instead an exercise in developing an estimate of a situation (this despite the fact that the wargame scenario was completely in line with the 1934 version of War Plan ORANGE).[2] His only concession to the operational use of the games was that they provided an opportunity for students to familiarize themselves with geographic areas of future strategic interest.[3] Kalbfus spent a large part of his tour personally revising the *Estimate of the Situation* booklet, turning it from a simple twenty-six-page outline of general rules in its 1929 edition to a ninety-three-page textbook titled *Sound Military Decision* by 1936. Kalbfus pushed hard to make the document even longer and more comprehensive, but his staff eventually persuaded him to accept a version that was more readable and useful to students.[4] This emphasis on theory versus practice was also reflected in another milestone for the college, the first session of the advanced course. This small class of admirals and senior captains who were already senior course graduates spent the year studying international relations, major strategy, and the broader aspects of warfare.[5] Unlike the junior class, the advanced class members did not participate in the wargames.

Like most of his predecessors, Admiral Kalbfus put his personal stamp on the college by reorganizing the staff. He arranged his instructors into two large departments. The Educational (later designated Operations) Department dealt with the classroom material and wargame execution, while the smaller Intelligence and Research (after 1936 referred to as simply Intelligence) Department expanded the role originally envisioned by Harris Laning by analyzing and critiquing all the chart and board maneuvers. This department also handled international law, student theses, and "strategic areas."[6] The staff maintained this organization for the remainder of the interwar period. Some of the more notable faculty members during the late phase included Captain Raymond Spruance, who returned to the staff in 1936 to teach tactics and eventually head the Operations Department until 1938. After his own graduation, Captain Richmond Kelly Turner '36, a staunch advocate of naval aviation and amphibious warfare, was Spruance's subordinate in the Tactics Department.

After the Van Auken report, War College strategic and tactical maneuvers entered a new phase of complexity in both format and content. The most obvious format change was the move into new facilities in Pringle Hall, which were much larger than the

FIG. 21. A photograph of a wargame in progress in Pringle Hall, circa 1954. This photograph shows to good advantage the grid squares on the game floor, curtains that shield parts of the board from view, and game equipment. The students in this postwar period wore their uniforms as opposed to civilian clothes. Courtesy of the Naval War College Museum Collection.

old maneuver rooms in Luce Hall, though maneuver scenarios did not always take advantage of the entire space. Photographs of games in progress in Pringle Hall such as figure 21 show that classes used only portions of the floor at a given time. But the important point was that maneuver space restrictions no longer limited game areas to what amounted to the visual horizon from a surface ship.

Figure 21, which probably dates from the mid-1950s, provides a useful comparison to figure 8 in chapter 2. The students are still using the same range wands, turning circles, and ship models that their predecessors used thirty years previously, but the scale on the game floor and the additional space accommodate much larger maneuvers. Figure 21 also shows the catwalk, installed to keep observers off the maneuver floor while allowing them a better view of game operations. A "cash carrier" pneumatic tube

U.S. NAVAL WAR COLLEGE, NEWPORT, R.I.   4 April 1934.
Contract NOy-1700 - View looking west showing Central Desk
in Maneuver Room.   No. NTS. "C" 34-112.

FIG. 22. A photograph of the Pringle Hall "cash carrier" message system, used to send messages between the plotting area and the individual rooms where the students played the roles of opposing fleet staffs. Courtesy of the Naval War College Museum Collection.

communication system (figure 22) linked the maneuver room with the planning cells of the opposing teams, replacing the Luce Hall messengers and contributing to the maneuver staff's ability to conduct games more efficiently. Pringle Hall would serve as the arena for tactical games until the introduction of computer-based wargaming in 1957.

As the classes progressed through the 1930s into the 1940s, the student and staff vision of naval war transitioned from a limited war for only naval objectives to total war that spanned a broad theater of operations. In terms of content, scenarios that encompassed three- to five-year periods became common in this phase, as did the inclusion of international partners or regional allies, notably the Netherlands East Indies (BROWN) and the Soviet Union (PURPLE).[7] Within those games, the students also maneuvered a series of smaller battles, which routinely included amphibious operations, as opposed to a single decisive fleet action.[8]

Unlike the other sample classes, the first class of the late phase was not dotted with future World War II leaders, though two faculty members bear mention. Captain Roscoe MacFall '24, Chester Nimitz's Naval Academy and War College classmate, was the head of the senior tactics section after three years on the faculty. As discussed in chapter 3, MacFall's peers knew him as a practical tactician, but he was also a theoretician with a gift for reducing the complexities of naval combat to simple dicta. He headlined his introductory remarks to the tactics course with the capitalized, underlined saying "The aim of strategy is concentration of purpose; the aim of tactics is concentration of force" and in the remarks emphasized the utmost importance of a clear estimate of the situation in achieving those aims. Reflecting an appreciation for naval history, MacFall's ideal in this respect was Nelson's "Memorandum before Trafalgar," widely considered the model of brevity and clarity in orders.[9] The second notable staff officer was Harry L. Pence '25, who headed the Intelligence Department and wrote all the maneuver critiques. Another instructor who frequently used historical references in his lectures, Pence continued Wilbur Van Auken's practice of writing detailed critiques and conclusions of the games. The student body was distinguished by a sizable junior class (forty-one members) that was almost as large as the senior class (forty-two). The two senior members of the class were *very* senior—Captains Charles S. Kerrick and Charles A. Blakely had already commanded major ships *before* their arrival at Newport.[10] Like William Halsey and Ernest King before him, Blakely was leaving the world of big-gunned surface ships for aviation. He finished the naval aviation observer course before arriving at Newport, and returned to Pensacola after his War College graduation to qualify as a pilot at age fifty-six.[11]

The class maneuvered nine STRAT, OP, and TAC problems, two RED and the rest ORANGE. As in classes before them, some STRAT or OP maneuvers flowed into TAC games. What distinguished the more advanced ORANGE games were going-in assumptions that the war had already been in progress for months, and the situation had matured sufficiently for BLUE to take the offensive from a base in the central Pacific as opposed to fighting its way across in one game. This is reflective of the cautionary strategy that gained provenance after the thruster strategy proved to be unrealistic in repeated wargames and fleet problems. The class also experimented with another new tactic in Tactical Problem II when they simulated the employment of mustard gas against both ships and shore bases. Army aviators in the class briefed the college staff and students on recent tests at the army center for chemical research at Edgewood Arsenal, where aircraft employed gas in tests against small boats. While the aviators disparaged the tactics because of the hazard to the dispensing planes, the class of 1935 used gas repeatedly in their games.[12] In his later years Carl Moore '35 recalled an exercise (most likely one of the seven quick-decision games played by the class) against his classmate Commander Mervyn S. Bennion in which Moore experimented with a gas attack and succeeded only in neutralizing his own ships.[13]

The Big Game for the class of 1935 was OP III, which flowed into Tactical IV. Admiral Kalbfus addressed the class on 20 February 1935, at the start of the maneuver, to stress that "mental exercise is, therefore, the prime motive and the medium [the game and scenario] is necessarily artificial." But in this same address Kalbfus also stressed the importance of students becoming familiar with what he termed the physical and strategic geography of the Pacific.[14]

The BLUE situation description for OP III-35 started with the assumption that the Asiatic Fleet and "the larger part of the white regular troops" stationed in the Philippines had been withdrawn upon the outbreak of hostilities, which of course removed the need for an immediate relief expedition. The months between this evacuation and the start of the problem took up three short paragraphs in the BLUE special situation description, which concluded with the BLUE battle and fleet marine forces safely based at Truk, with the Caroline and Marshall Islands under U.S. Army control. The BLUE commander's task was to sortie to the northwest, engage the ORANGE fleet and seize the island of Guam with an amphibious assault.[15] The assault executed in the chart maneuver was a rudimentary affair, involving the landing of twelve marine and army battalions in ideal surf and weather

The Late Phase

conditions over beaches that were virtually undefended. The only significant casualties suffered by the BLUE landing force were from gas. The writer of the "Comments on Landing Operations at Guam" enclosure to the postgame analysis was probably marine colonel Richard Cutts, the only U.S. Marine Corps officer on the faculty that year. He criticized the BLUE landing off-load plan as overly time-consuming and noted that the lack of defenses at the landing site was unrealistic. But while these comments showed that the Guam landing operation could have been better planned and executed, the specific points of the comments indicate that the class of 1935 played out a land combat situation to a level of detail not seen previously in a Naval War College wargame.[16]

Through the rest of the strategical phase of the maneuver, the BLUE fleet, reinforced with auxiliary aircraft carriers, ranged over the central Pacific, not only invading Guam but also conducting air strikes against Manila, Saipan, and the "Pelews" (Palau) while ORANGE put up minimal resistance. In the postgame critique, Harry Pence took ORANGE force commander Captain Kerrick to task for lack of aggressive employment of his numerically inferior forces. Apparently, Kerrick elected to restrict his operations to attrition attacks and conserve his strength to engage BLUE close to the Philippines during the tactical phase. This indicates, and the critique confirms, an understanding on the part of Kerrick and the rest of his ORANGE staff that no matter what the BLUE fleet accomplished in the central Pacific they were eventually going to have to come west and fight ORANGE in the Philippines.[17]

The tactical phase of the game was set in the Celebes Sea south of the Philippine archipelago, between the early and middle phase objectives of Dumanquillas and Tawi Tawi. The large junior class joined the senior class in executing the maneuver. Captains Blakely and Richard S. Edwards filled the respective roles of BLUE and ORANGE commanders in chief. The missions assigned to the opposing fleets were succinct—to seek out and destroy their opponents and to gain or maintain control of vital sea areas. In a throwback to early phase tactics, instructions to both BLUE and ORANGE fleet elements gave priority to reducing the fighting power of their opponent's battle line.[18] The fleets did depart from the early phase practice of symmetrical formations; for example, the ORANGE approach and battle dispositions had destroyer and cruiser divisions arranged in a V with eighteen thousand yards (nine nautical miles) of separation between the battleships at the base of the V and the light forces in the forward legs.[19]

The resultant melee between opposing battle lines did not differ greatly from tactical maneuvers fought in the early and middle phases. The annotations in the maneuver blueprint records show that both BLUE and ORANGE launched large air attacks that inflicted significant damage to ships on each side, but the primary weapons employed were destroyer-launched torpedoes. Figure 23 captures moves 20 and 21, with the ORANGE battle line of six battleships (BBs) and battle cruisers (CCs) to the west and the BLUE battle line of seven battleships to the east. The dotted fan-shaped markings in the center of the chart between the opposing battle lines indicate torpedo attacks, and the notations next to the fans (e.g., 5T) indicate the number of torpedoes fired in each "spread." In the time encompassed by this chart, there are at least 320 torpedoes in the water. The postgame analysis of the tactical phase recorded that ORANGE ships fired 748 torpedoes and BLUE fired 430.[20] The high volume of fires in such a short period gives an indication of the challenges inherent in capturing important data from the games, why the games took as long as they did, and the lengths to which the staff had to go to overcome those challenges.

One major aspect that set OP III-35 apart from previous games was the volume of quantitative data produced, and how the faculty used it to interpret the game outcome. The Intelligence Department captured the rapid reduction in BLUE and ORANGE fighting power in composite metrics of "fire effect" (a function of guns remaining) and ship "life," as well as damage scores for each capital ship. Figure 24, a re-creation of a chart from the blueprint record of moves from OP III-35, shows the total capital ship damage plot for each fleet for OP III and reflects the larger

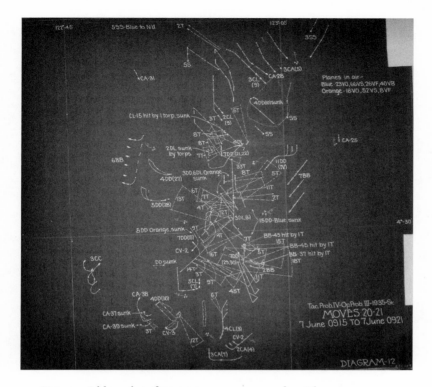

FIG. 23. A blueprint of TAC IV-35 moves 20 and 21. These moves are notable for the large number of torpedoes being fired, indicated by the (#)T markers. U.S. Naval War College, "Operations Problem III-1935; Blueprint Record of Moves, Diagram 12," Folder 1914-1, Box 70, RG4, Publications, 1915-77, Naval Historical Collection, Naval War College, Newport RI.

BLUE battle line clearly absorbing more damage. For example, BLUE battleships received twelve hits in the fusillade of torpedoes fired between moves 20 and 25, which resulted in three sunk and another put out of action.

Table 3, another re-creation from OP III-35 records, summarizes the state of both BLUE and ORANGE battleships at the end of the game. BLUE has less life remaining but more fire effect than ORANGE, which reflects the greater number of heavy caliber guns on American battleships than on their Japanese counterparts. While the emphasis on capital ship damage as a measure of effectiveness is archaic, these attempts at quantifying overall results were new to the games.

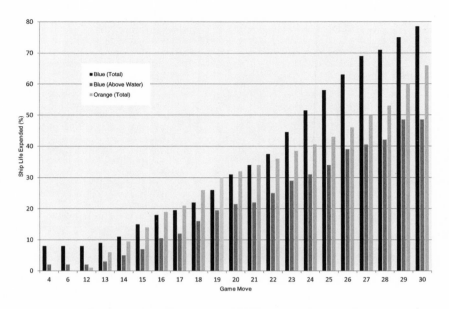

FIG. 24. TAC IV-35 capital ship damage. This chart shows the cumulative damage absorbed by the BLUE and ORANGE capital ships during the TAC IV-35 maneuver between moves 4 and 30. BLUE's damage exceeds that inflicted on ORANGE, and the majority of BLUE's damage came from ORANGE gunfire (above-water damage) as opposed to torpedoes. U.S. Naval War College, "Operations Problem III-1935; Blueprint Record of Moves, Diagram 12," Folder 1914-I, Box 70, RG4, Publications, 1915-77, Naval Historical Collection, Naval War College, Newport RI.

Besides featuring highly detailed game results, the postgame critique for OP III / TAC IV was notable for discussion of the tradeoffs that were inherent in the cautionary strategy. Captain Pence noted that by proceeding slowly BLUE "telegraphed his objective" while gaining the consolidation and security of his lines of communication in return. On the other hand, the critique presaged the Guadalcanal campaign of 1942 by noting that the ORANGE strategy of attrition could succeed only if the attacker was able to inflict greater losses than he absorbed.[21] The critique is also important in what it does not include, which in this case is the usual extensive discussions about the practicality of air attack during the strategic phase. The absence of this discussion reflects a certain degree of acceptance on the part of both staff

The Late Phase

Table 3. TAC IV-35, capital ship damage at the end of move 30

Blue ships remaining			
Damage	*Above-water (%)*	*Underwater (%)*	*Total (%)*
BB 45	29.5	29.2	58.7
BB 44	29.4	16.9	46.3
BB 42	38.4	48.5	86.9
BB 41	37.1	34.7	71.8
BB 39	49.6	10.5	60.1
BB 37	53.5	27.3	80.8
BB 35	37.2	47.3	84.5
Sunk	BB 46, 43, 38		
	BB 35 no fire effect—list		
Total fire effect remaining	50.20% (6 ships)		
Total life remaining	21.50%		
Orange ships remaining			
Damage	*Above-water (%)*	*Underwater (%)*	*Total (%)*
BB 5	47.6		47.6
BB 4	37.8		37.8
BB 3	45.9		45.9
BB 2	65.3		65.3
BB 1	75.7	18.5	94.2
CC 3	44.5		44.5
CC 1	71.6		71.6
Sunk	BB 6, CC 4		
	BB 1 no fire effect—sinking		
Total fire effect remaining	37.00% (6 ships)		
Total life remaining	33.50%		

A table showing capital ship (battleships [BB] and battle cruisers [CC]) damage at the end of move 30 in the TAC IV-35 maneuver. Higher numbers indicate greater damage. This table was created from the data shown in charts that were part of the original maneuver results. *Source*: Department of Intelligence, U.S. Naval War College, "Abstract of Trans-Pacific Problem, Beginning 20 March 1935; OP. Prob III-Tac. Prob IV-1935-SR; Tactical Phase," Folder 1914-H, Box 70, RG 4, Publications, 1915–77, NHC.

and students of the place of naval aviation in the naval battle. Notes taken at the critique illustrate the give-and-take between staff and students, and show that members of the junior class were actively involved and even encouraged to provide insights. Lieutenant Charles F. M. S. Quinby, one of the most junior members of the junior class, provided the combined classes with information on sonar performance that he received while attending the "super-sonic" (sonar) school, demonstrating knowledge in an important warfare area that his more senior classmates did not possess.[22]

After the class of 1935 graduated, Admiral Kalbfus continued to update and expand *The Estimate of the Situation*. This effort remained his primary focus until the end of his tour, and he appears to have had less interaction with wargaming during this time. In December of 1936, Kalbfus was relieved by Admiral Charles P. Snyder '25, who had served a short tour as a member of the Operations Department staff under William Pratt. During these years, the effects of fleet expansion became evident. The junior class decreased precipitously in size from a high of forty-one in 1935 to an average of sixteen between 1936 and 1940, probably due to the increased need for junior officers in the expanding fleet. The senior class size stayed constant and robust, however, maintaining an average of fifty students during the same years.[23]

In 1935 Captain Raymond Spruance joined the staff for his second tour as an instructor, this time replacing Captain Milo Draemel '26 as head of tactics for the junior class. Draemel, a thoughtful strategist and planner, had provided considerable continuity to the faculty, as he served in the Operations or Educational Departments from 1928 to 1931 and again from 1933 to 1936.[24] Spruance moved to the senior class position the next year, and then up to head of the Department of Operations in 1937. Kelly Turner '36 took Spruance's position as senior class tactics head that year. Spruance, Turner, and other new members of the Operations Department staff were the prime movers behind a comprehensive modernization of the War College curriculum. Bernhard Bieri credited Turner in particular with direct-

The Late Phase

ing instructors to become much more familiar with the material they taught, and to be able to conduct their lectures without the aid of scripts.[25] Lecture notes and summary records from the later phase also reflect much more emphasis on aviation, submarines, and amphibious operations than in previous years. The scenarios developed by Turner and the rest of the operations staff began to branch out beyond the usual RED and ORANGE situations. The class of 1937 played their TAC II maneuver off the West Coast against a hypothetical WHITE opponent. WHITE was similar to ORANGE but possessed greater strength.[26] Allied nations started to factor into the games to a much greater degree than in previous years. PURPLE (the Soviet Union) appeared again as a BLUE ally in OP III-37, setting the stage for a multilateral game in 1938.

The class of 1938 contained the largest number of future flag officers of all the interwar period classes. They included Captain Robert L. Ghormley, who as a vice admiral five years later would lead the initial months of the Guadalcanal campaign; aviation pioneer Captain Aubrey Fitch; and Commander Charles "Soc" McMorris, a surface officer already recognized by his peers for his keen intellect who would eventually become Fleet Admiral Nimitz's plans officer during the war.[27]

OP VII-38 was the major trans-Pacific maneuver for the class of 1938 and featured the greatest degree of geopolitical influence of all wargames to date. The situation description depicted BLUE in a secret alliance with BROWN (the Netherlands East Indies), while SILVER (Italy) and BLACK (Germany) made their first appearances in a War College game as allies of ORANGE. SILVER and BLACK did not figure directly in the game, though the scenario credited them with keeping PURPLE from intervening in the Pacific theater. On the other hand, BROWN appeared in this game as an active participant with the roles of BROWN force commanders played by a team of five students led by Captain Ghormley. The statement of the problem for OP VII-38 also mentioned political pressure for decisive action due to BLUE home front impatience with what was already a two-year war, as well as economic pressures on ORANGE. ORANGE was also handicapped by a need to avoid the "danger of seriously offending RED and GOLD (France)."[28] From the point of view of the fleet commanders, these political conditions simply hurried both forces to action, but they also gave the students at least some idea of the complexities inherent in coordinating military actions with allied nations.[29] In this respect the staff's inclusion of BROWN in the order of battle was prescient, but what the staff failed to anticipate was that none of the real nations symbolized by PURPLE, BROWN, RED, or GOLD would be in any position at all to influence events in the Pacific in 1941.

The class of 1938 played the strategic chart maneuver portion of OP VII-38 from 7 to 23 April and the tactical board maneuver portion from 25 April to 2 May. The critique ran from 9 to 11 May, which made this game the next to last maneuver in the class curriculum.[30] One of the more striking aspects of this game was the geographic spread of the chart maneuver. Fleet elements engaged in the central Pacific between the Mandates and the Philippines, and in the eastern and central parts of the Netherlands East Indies. ORANGE forces attacked BROWN at Makassar on the island of Celebes, while the BLUE forces conducted an invasion of the ORANGE-held islands of Buru and Ceram seven hundred miles to the east.

The staff solution for BLUE assigned substantial tasks to the small BROWN fleet, charging it with holding off ORANGE thrusts against Java, Sumatra, Celebes, and Borneo until BLUE forces could secure the islands of Ceram and Buru in the eastern end of the theater. The intent behind these moves was to clear a route to Java that would allow BLUE to establish a base there to support an eventual invasion of the Philippines.[31] Figure 25 illustrates part of move 6 of the game, the BLUE fleet approach to Ceram (now Seram). Ceram is the island at the bottom center of the chart. The chart itself shows much less detail with regard to fleet formations than records of previous years. Positions of the individual BLUE forces, the battle force, base attack force, support force, and train appear as simple squares as opposed to the precise geometric formations seen in the records of early phase maneuvers. The position and track markings surrounding the BLUE force reflect how students could simulate the scouting

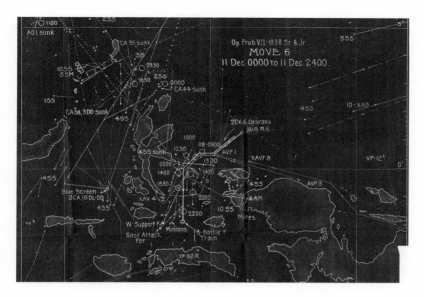

FIG. 25. A blueprint of move 6 in OP VII 38, a maneuver played by both the junior and senior classes. At this point in the maneuver, the BLUE invasion force is approaching the island of Ceram (the land mass in the bottom center of the blueprint) in what was then the Netherlands East Indies while the BLUE screening force is engaging ORANGE to the northwest. U.S. Naval War College, "Operations Problem VII-1938; Blueprint Record of Maneuvers, Move 6," Folder 2166M, Box 84, RG4, Publications, 1915-77, NHC.

operations of long-range reconnaissance aircraft (indicated by the VP label) and submarines (SS) without the artificial restrictions of game board size.[32]

The chart illustrates the screened approach of an invasion force toward enemy-held territory as opposed to a sortie to find and engage the opposing fleet. This recognition of what Michael Vlahos called "strategic geography" indicates the greater influence of Kelly Turner, who by this time had moved from the tactics section to the head of strategy for the senior class. According to Bernhard Bieri, Turner orchestrated a major overhaul of maneuver problems to "bring them into line with the most probable developments in naval expansion."[33] The new emphasis on amphibious operations especially reflected the impact of these updates. The records of OP VII-38 also contained detailed plans for the ORANGE defense and BLUE invasion of Jolo, a small but strate-

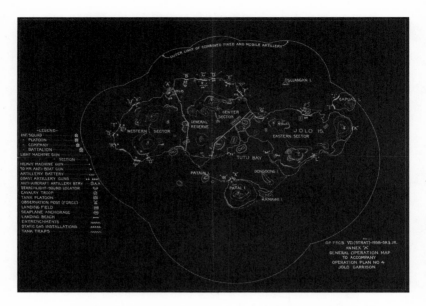

FIG. 26. A blueprint showing the ORANGE garrison and defenses on the Philippine island of Jolo during the OP VII 38 maneuver. This diagram reflects the greater emphasis paid to amphibious operations in maneuvers of the late phase. U.S. Naval War College, "Op. Prob. VII (Strat) 1938-SR & JR; Annex 'A,' General Operation Map to Accompany Operation Plan No. 4; Jolo Garrison," Folder 2166-M, Box 84, RG4, Publications, 1915-77, NHC.

gically placed Philippine island south of Mindanao. Jolo was well to the northwest of the OP VII center of action and did not figure in the maneuver itself, so it is probable that these plans were either part of an illustrative staff solution or a student committee exercise. Figure 26 is a plan for the ORANGE defense of Jolo and shows details down to gun emplacements and troop dispositions.

Figure 27 illustrates the gunfire support plan for the BLUE assault on Jolo and provides an interesting comparison to the early phase landing communications plan for OP III-27. While the 1927 plan was highly detailed, it focused on communications equipment and not so much on the troop movements that the communications were intended to support. Advances in radio reliability by 1938 relieved the need to lay extensive cable systems, and the OP VII-38 planners could focus more on coordinating gunfire support with the anticipated movements of the landing

The Late Phase

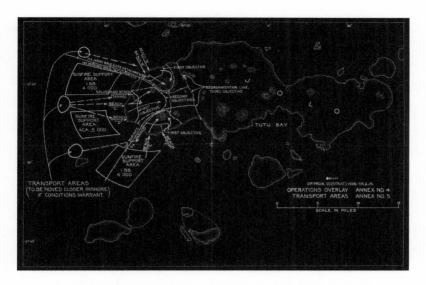

FIG. 27. A blueprint showing the BLUE gunfire support plan for an invasion of the ORANGE-occupied island of Jolo during the OP VII maneuver for the class of 1938. U.S. Naval War College, "Op. Prob. VII (Strat) 1938-SR & JR; Operations Overlay Annex No. 4, Transport Areas Annex No. 5," Folder 2166-M, Box 84, RG4, Publications, 1915-77, NHC.

force. This sort of close coordination between sea-based fire support and land forces would become a major feature of American expeditionary operations during the coming war, and it is worth recalling that none of the officers who executed these first operations during World War II had ever conducted a live one previously. Their only exposure to amphibious assault planning prior to actual combat was at the Naval War College. The 1936 War College lecture on the World War I Gallipoli campaign particularly emphasized the need to devise a system of naval gunfire support, and the Jolo charts reflect War College efforts to provide students with some familiarization with this discipline. Both the Jolo plans and the description of the BROWN defense of Makassar also show a level of attention to amphibious and ground operations far above the offhanded discussion of the Guam invasion maneuvered in OP III-35 just three years prior.[34]

The class of 1938 conducted one more tactical maneuver before graduation. OP VIII-38 involved a BLUE fleet "retirement" from

Java to its base in Staring Bay under pressure from a stronger ORANGE fleet. The scenario ruled that bad weather had grounded all shore-based aircraft, and no aircraft carriers were included in the order of battle for either side. This allowed the game to commence when a sudden clearing in the weather revealed the opposing battle lines to each other.[35] The resulting situation was much like a prolonged quick-decision maneuver. Like TAC IV-35, OP VIII-38 featured heavy use of torpedoes, with the BLUE fleet firing a total of 246. When compared with other games of the late phase, the OP VIII emphasis on surface attacks seems only something of a tactical retrograde, but the postcritique was still notable for its detailed calculation of damages. Figure 28 illustrates a compilation of significant ship damages (the original damage charts spanned five separate pages), and reflects how the collection and display of maneuver results development had matured since the early phase. Damage recorded in this chart is not limited to capital ships, as was the case with TAC IV-35, but also includes cruisers and destroyers.

If the middle phase was a coming-of-age for the War College and its wargaming program, the last six years of the interwar period were a time of steadily increasing maturity and relevance. Game records from the period show that by the middle of the 1930s the students and staff were simulating a Pacific war very much like the one they ended up fighting in the 1940s. Staff and faculty members who were supporters of emerging naval doctrines such as aviation, expeditionary logistics, and amphibious warfare were no longer working on the fringes of game development but were instead operating in the center and influencing the career development of officers who would be commanding ships and squadrons in the coming conflict. Forward-thinking advocates like Kelly Turner could demonstrate these capabilities in what was by now a proven and widely accepted venue in the wargames. Battle lines of battleships were not completely displaced from the center of naval tactics and strategy, but they had certainly been compelled to share the space.

By the time that the class of 1938 graduated, the hostile peace of the late phase was giving way to preparations for open combat.

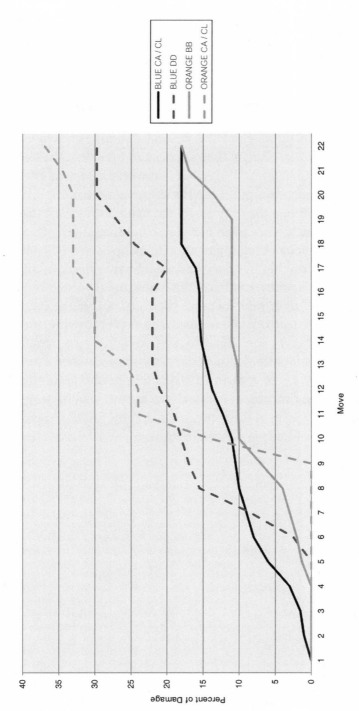

FIG. 28. A chart showing the cumulative damage absorbed by BLUE heavy and light cruisers (CAs and CLs) and destroyers (DDs), and by ORANGE battleships (BBs) and cruisers during the OP VIII 38 maneuver. U.S. Naval War College, "Operations Problem VIII (Tactical) 1938; Damage Record Moves 1–22," Folder 216/D, Box 84, RG4, Publications, 1915–77, NHC. The reduction in BLUE damage that appears between moves 16 and 17 could reflect battle damage repair but is probably a data-recording error.

On 12 December 1937, while members of the class were engaged in developing their solutions for one of the more basic operational problems, Japanese aircraft attacked and sank the U.S. Navy gunboat *Panay* (PR-5) on the Yangtze River. During March of 1938, when the class started OP VII, Germany annexed Austria. Five days after the class graduated, Congress passed the second Vinson-Trammel Naval Expansion Act, also known as the "Two-Ocean Navy Act."[36] Two years later, President Franklin Roosevelt would state, "Today our Navy is at a peak of efficiency and fighting strength. Ship for ship, man for man, it is as powerful and efficient as any single navy that ever sailed the seas in history. But it is not as powerful as combinations of other navies that might be put together in an attack upon us."[37] Given the huge expansion that the navy had seen by that time, Roosevelt's hyperbole is understandable. But to say that the U.S. Navy was "ship for ship" and "man for man" the equal of Japan or even Great Britain was to be mistaken. Given the slowly growing recognition that the navy was going to have to fight a two-ocean war against a combination of foes, the comparison looked even worse. Soon the class of 1938 and all the other graduates of the War College from the interwar period would be at war, fighting against real opponents in the same areas that they had marked out on the floors of Luce and Pringle Halls.

Conclusion

The interwar period ended with the U.S. Navy on the precipice of war, a war that naval leaders had attempted to predict and practice for twenty-two years in fleet problems and wargames. However, to say that the navy *practiced* the Pacific war is not to say that they predicted it with complete accuracy or were totally prepared for it when it came, as Admiral Nimitz's statement quoted at the beginning of this book would seem to imply. Taken at face value, the Nimitz quote is an oversimplification or at least a distracter. Behind the words, though, is a claim that the trained officers the War College fed into war-planning positions played a significant role in transforming the U.S. Navy, with its post–World War I physical state and mindset, into a military force that was much better prepared to fight a real war with Japan. This book explored the question of what roles the interwar period wargames at the War College played in that transformation and to what extent any military organization could transform itself, even with untested principles or unproven technology.

This question is really one of agency and the instruments of agency. Historical agency is the ability of an entity to influence the development of the events of history. By that definition, the question becomes whether the Naval War College was a historical agent of preparation, transformation, and innovation in the same way that Kuehn argued for the General Board, and whether the wargames were instruments of that agency in the same way that Felker and Nofi argued for the fleet problems. The answer is contained in the War College wargame records of the inter-war period. They show that, far from being irrelevant rituals, the wargames were definitive instruments of agency. Not only that, but the records reflect that the wargames were an *effective* instrument, at least within the limits of their stated objective. Through lectures, readings, and especially wargames, the War College taught decision making and not decisions. The decisions and results were important derivatives of the War College experience, as they gave the student-officers an adaptable process to follow and confidence in their decision-making abilities.

While he was conducting the research that led to his biography of Raymond Spruance, Thomas Buell came to some of his own conclusions that touched on the subject of interwar period transformation. He stated that on a *strategic* level, the wargames only partially prepared naval officers intellectually and psychologically for the war before they had to fight it. His "partial" qualification rested on the fact that none of the classes ever played the Battle of the Atlantic on the game board, which is true enough. The RED games in the greater part of the interwar period and even the BLACK-SILVER games of the final years were still fleet-on-fleet actions and not the war of attrition between escorted convoys and U-boat wolf packs that the Battle of the Atlantic turned out to be. Buell's assertion was that the major changes in the navy brought on by the wargames were on a *tactical* level.[1]

The evidence reviewed in this book supports a different view, that the games did facilitate naval transformation across not only tactics but strategy and technology as well. That transformation was far from complete in December 1941, and the list of unpleasant surprises for the U.S. Navy in that war went far beyond kamikaze attacks, but the ability to repeatedly practice procedures and experiment with innovations in a low-cost, flexible venue gave the wargames a central role in that transformation.

Basil Liddell Hart defined strategy as "the art of distributing military means to fulfill the ends of policy."[2] Roscoe MacFall condensed that definition into "the concentration of purpose." By either one of these definitions, the navy certainly did change strategically and the games had a definite influence on this change.

The conclusions and recommendations of the 1933 Van Auken report, reinforced by the experiences of the scores of students who played the games he documented and then moved on to OP-12, was a deciding factor in the shift from the Mahanian thruster strategy to the more realistic cautionary strategy. One of the most striking reflections of this departure from the early-phase Mahanian doctrine occurred in the first months of the Pacific war. When the Pacific Fleet reinforced and reconstituted itself shortly after the Pearl Harbor attack, a battle force expedition to relieve MacArthur's army in the Philippines—the bedrock rationale for the thruster strategy—was never seriously contemplated.[3] The island territories of Guam and Wake were similarly written off. Surviving, gaining battle experience, and building up for a prolonged war of attrition and a step-by-step advance comprised the navy's early Pacific strategy.

The wargames also added to a growing realization that a Pacific war would necessarily involve other nations besides ORANGE and BLUE. Wargame critiques from the middle phase frequently stressed the necessity and challenges of establishing forward logistics bases. While it is true that the bulk of naval fighting forces in the Pacific theater were American, their recognition that they needed to push logistics support forward and their ability to use bases in Australia, New Zealand, Vanuatu, and French Polynesia greatly relieved logistical difficulties during the Solomons and New Guinea campaigns.

Tactically, the navy departed dramatically from the battleship battle lines that figured so prominently in early and middle phase wargames. For the first six months of the war, task forces built around aircraft carriers and cruisers made small forays against Japanese bases in the Mandate islands, but the intact or slightly damaged battleships—not only slow but also voracious consumers of fuel in a fleet that was critically short of fleet tankers—stayed close to home.[4] Later, in 1942, when Nimitz arrayed his forces to face what he knew was a numerically superior opponent at Midway, he had the option of bringing seven battleships west to augment his two carrier task forces. Nimitz made a conscious decision to leave his battle line on the sidelines, though he had over twice as many battleships available as he had carriers at his disposal. His official rationale was "the undesirability of diverting to [the battle line] screen any units which could add to our long-range striking power against the enemy carriers."[5] In other words, Nimitz viewed the battleships more as a defensive liability than a contribution to the "striking power" of his force. His intent was to engage the Japanese fleet with his carriers alone. Certainly, Nimitz did not come to this decision overnight or even in the immediate aftermath of Pearl Harbor. The War College wargames and fleet problems had been illustrating the vulnerabilities of a battle force built around battleships for years. Increasing awareness of the potential of aviation is evident from as far back as the class of 1923 in statements from Harris Laning and others, and had ceased being a point of contention by the middle of the late phase. Nimitz's decision, made only seven months into the war, stands at odds with the popular canard that all senior navy leadership at the time of Pearl Harbor remained myopically focused on battleships.

On the other hand, the wargame process appears, in retrospect, to have been a poor venue to experiment with and develop submarine tactics. The causes of this failure are closely related to systemic problems in the submarine force that did not make themselves apparent until the start of hostilities. An excellent source for the complete history of the submarine service in World War II is Clay Blair's *Silent Victory*, but in summary, the deficiencies covered the full spectrum from strategy to tactics, and especially technology. Strategically, submarines were still something of an unknown quantity in interwar period navies. Only Germany had any significant experience in a submarine campaign, and Great Britain was still making attempts as late as 1930 to ban submarines altogether. The London Naval Treaty, which required submarines to abide by prize rules, outlawed unrestricted submarine warfare against commercial shipping. In the interwar period, the United States and other navies experimented with "submarine cruisers" equipped with large-caliber guns and spotting aircraft (such as the French *Surcouf* and British *M2* and *X1*) and in a mine-laying role (such as USS *Argonaut*), reflecting the conflicting views of how they should be employed. U.S. Navy submarine tactics reflected this strategic confusion. Since the War College emphasized combatant actions, especially fleet or at least task force engagements, submarines were assigned to screening duties on the periphery of surface ship formations. Submarine tactics emphasized caution, avoiding detection, and submerged sonar approaches to targets.

U.S submarines were most deficient in the technical sense. Shortcomings in habitability, seakeeping, and engine reliability were well known during the interwar period, but crippling deficiencies in torpedoes did not reveal themselves until after the start of hostilities. Due to small budgets and service infighting, U.S. Navy torpedoes were inadequately tested, but the small sample of test results was sufficient to convince Navy leadership that their torpedoes would work as advertised under operational conditions. These assumptions were reflected in maneuver rules and in War College wargames. Out of favor politically, chronically underfunded, and lagging far behind aviation and expeditionary warfare in terms of emphasis, the U.S. submarine force had to catch up and practice under combat conditions. This process took years, at the price of scores of unsuccessful patrols and avoidable losses. Once overcautious commanders were replaced with more aggressive officers, proper tactics were developed, and, most of all, torpedo deficiencies were diagnosed and corrected, the submarine force rapidly evolved into a major factor in the eventual defeat of Japan.

Surface tactics were the type most often exercised in the wargames, and the applicability of the lessons learned in those games to real combat, when it came, is questionable. Students moved from rigid linear formations to more flexible circular approach formations in the early phase, and then relaxed the need to remain in precise geometric formations during the late phase. The Solomons campaign of 1942–43 encompassed the greatest number of surface engagements fought during the war, but two things that the wargame designers did not foresee (and that Admiral Nimitz did not mention in his speeches) were the Japanese capabilities in night combat and the superiority of their

Long Lance torpedoes, which drove the outcomes of those individual battles.

The wargames also provided a venue for students to experiment with some of the new or proposed technological developments of the interwar period. Rigid airships carrying aircraft, gas attacks against land and sea targets, and cruisers with flight decks were all tested, and their mixed performance as reflected in game results undoubtedly played some part in the navy's decisions not to continue with them. On the other hand, developments such as floating dry docks, aircraft carriers converted from merchant ships, conversion of older combatants for fire support roles, and recommendations for design changes to submarines and for antisubmarine warfare sloops were continued from ideas to design, construction, and deployment.

The game was a constant presence in the interwar period, but it was not a solid, tangible entity that the War College could box, label, and place on a shelf like an Avalon Hill board game. Neither did it resemble the hardware, software, and documentation of the Naval Electronic Warfare Simulator. The physical components of the wargames were simply a series of rooms, a stack of manuals, and some basic measuring and drawing equipment. Neither was the game an oracle or a crystal ball that provided a view into the future. The literal beating heart of the games was that of the people who worked in and around them. These included farsighted War College presidents like William Sims, William Pratt, and Harris Laning, who created and sustained an environment that encouraged and nurtured the initiative of faculty members like Raymond Spruance, Kelly Turner, and John Greenslade. These men in turn constantly updated maneuver problem scenarios, encouraged innovation, challenged their students to explore original solutions, documented their lessons learned, and passed them down from year to year.

One factor that contributed to making the games as useful as they were was that the staff and faculty of the college always considered them as primarily educational tools. There is no evidence in the statements of problems, postgame critiques, or player memoirs that suggests there were any scripts, agendas, or specific programs being showcased, as is often the case today. Students played on both sides, and played their best games regardless of their assigned role. These were not pro forma games, and there is no primary source evidence that results were ever varnished or "spun" to favor BLUE. One might point to John Hattendorf's anecdote about Captain Ernest King's preference for a northern approach route for his BLUE fleet during OP IV-33 as opposed to the staff's recommendation for a southern route, but the wargame critique shows this to be an attempt to move away from a strategy already proved to be unworkable to one not yet attempted.

Another cause of wargame influence was their sheer ubiquity. Compared with what it would become after World War II, the officer community of the interwar period was small, and the War College was a common tour of duty for officers on their way to senior ranks. Officers attended the school, cycled back out to the fleet, and then returned for additional assignments as instructors. Almost all the officers listed on the registry as staff and faculty were previously students, and six of the eight interwar period War College presidents were former instructors or staff members. Student-officers also played the wargames in a real-world environment with generally current orders of battle as opposed to notional or fictitious scenarios set years into the future. The upshot of this rotation of officers was a continual communication between the fleet and the school, which allowed graduates to more readily compare what they experienced in real-world operations with what they saw on the game floor. With this rotational system in place, the game was continuously updated and refreshed. Wargaming was a widely shared experience among the senior officer corps, and War College methods and lessons became pervasive throughout the navy. Vice Admiral Olaf Hustvedt '41 was specific in his postwar assessment of the relevance of his War College training. In an interview with the U.S. Naval Institute Oral History program, Hustvedt recalled:

A couple of years later I was in the Pacific when the attack on the Marshall Islands took place, and when we attacked Truk. . . . [T]hat brought Admiral Ike Giffen and me together for the first time since we had been neighbors at the War College, and . . . after the immediate fracas around Truk was over . . . I had time to send a little PVT, private message to Admiral Giffen on his flagship. I said something to the effect, "How are you, Ike? It's great to meet up again on the old campus," because we were actually operating around Truk which we had done on the game board at the War College a year or two before![6]

Hustvedt is referring to OP 11-41, the BLUE defense of Truk against an attacking ORANGE force. Anecdotal evidence suggests that during the war, navy planners painted a map of the Pacific area of responsibility on the expansive concrete lanai floor of the Pearl Harbor Submarine Base bachelor officers' quarters—adjacent to Admiral Nimitz's Pacific Fleet headquarters—and used it for wargames while marines guarded the building entrances. While this story is anecdotal, it is not implausible. Most of Nimitz's senior staff were War College graduates.

Finally, the War College environment also fostered a quantitative approach to measuring the results of naval tests and experiments, an approach that other venues such as the fleet problems could not replicate. The games were a comparatively data-rich source, due to successive iterations of similar games. The actual numbers of games conducted during the interwar period is open to question. The records are not complete, and some games such as demonstrations and quick-decision problems do not belong in the same category as the major trans-Pacific problems. But the true number of games is much less important than the fact that for the twenty-two years of the period the college conducted four to six two-sided games annually, which provided a significant data set by any measure.[7] In fact, these two wargame attributes— that they were ubiquitous and quantitative—provide an argument that the most significant part of the Nimitz quote was not his reference to the surprises but his use of the phrase "so many people and in so many different ways." Virtually the entire U.S. Navy officer corps had been preparing to fight the Japanese in the Pacific for the whole interwar period in one venue or another, and most of the senior leadership of the navy had done so at the Naval War College.

In the final analysis, the story of U.S. Navy preparation for World War II is not about the fleet problems, the wargames collectively, or any single game. Individual games were "simply vehicles for the transportation of ideas from the abstract to the concrete."[8] The story is more about how the maneuver problems were continuously repeated—differing in detail but constant in theme—and the number of students exposed to them. The interwar navy was a tight, professional community, and the War College games were a shared experience of virtually all naval leaders. The very similar situations were played every year with different students, many of whom came back to the school as instructors, bringing with them a balance of theoretical and practical knowledge. The games were not innovations in themselves. Instead, they were a common playing field, a shared experience, a flexible constant, and a proving ground. The games were transformative because the staff and faculty who administered them recognized their educational role and remained adaptable to changing conditions. The student of 1923 would have recognized the mechanics of the games of 1936—maybe not the scenario or the ships, but certainly the game experience. Like Sims's and Laning's football metaphor, the players changed but the game did not.

Appendix A

Naval War College Class Demographics

The composition of every class at the War College between 1919 and 1941, including student rank, service, and branch where applicable and identifiable, is described in this appendix. Its intent is to quantify and trace the evolving demographics of those officers assigned to the War College over the interwar period. The data to support this section was extracted from the Naval War College's *Register of Officers, 1884–1955*.

The Naval War College frequently represented the only service schooling available for naval officers during the interwar period and for a long time afterward. The only other naval service school, the Naval Postgraduate School, focused on advanced engineering during this era, and general line officers did not typically attend it. By contrast, the U.S. Army established a formal progression of professional schooling for its officers. This progression started with basic branch (infantry, artillery, et cetera) school for newly commissioned officers, continued to advanced branch schools for experienced company-grade officers (first lieutenants and captains), then went on to the Command and General Staff School for field-grade officers (majors and some lieutenant colonels), and ended with the Army War College.[1] Army selection boards determined who would attend each of these schools. But while the army schools were more prevalent, they were also much narrower in focus. In terms of equivalence, the interwar period Naval War College fell between the Command and General Staff School and the Army War College.

The officer corps of the U.S. Navy declined dramatically after the end of World War I to a low of 7,855 in 1922. From this point the population of officers remained in the 8,000-plus range for the rest of the 1920s and in the 9,000-plus range in the 1930s until wartime expansion took hold in 1939. Despite this reduction in size, course offerings at the early phase War College expanded and annual enrollment steadily increased. The original six-month course grew to a full year term in 1922. When the junior course for unit command preparation for lieutenant commanders and below commenced in 1924, the original course of instruction then became the senior course for commanders and above. It continued the original War College mission of training officers for fleet staff duty and major commands.

Class enrollment increased from thirty in 1920 to an early phase high of seventy-one (forty-six senior and twenty-five junior) students in 1926. During the first years of the phase, classes consisted of a mix of ranks from junior lieutenants to captains, with a few admirals scattered throughout the class rosters. Navy officers made up the bulk of the student body, with line officers (surface line and, later, naval aviators) making up the majority of the navy representation. A small number of navy officers from the staff corps (medical, supply, and civil engineers) also attended. Once the junior course was established, navy commanders made up the largest population in terms of rank in the senior course. In the interwar period, a commander at the War College generally had somewhere in the area of twenty years of service.[2] The army and marine corps were regularly represented, with three to four officers from each of these services attending every senior class. During the early phase, most army student-officers came from the coast artillery corps, a branch whose coastal defense mission put them in frequent contact with the navy. One or two coast guard officers also attended each session until 1925, when the mounting pressures of Prohibition enforcement required an increased operating tempo at sea.[3] Table 4 lists the demographics of the early phase War College classes.

The middle phase years saw the junior class grow to nearly equal and in 1929 exceed the senior class size. Navy line commanders and captains continued to make up the bulk of the senior classes, with navy line lieutenant commanders and lieutenants comprising the great majority of junior classes. Attendance by other service students increased to a point: the army and marine corps continued to send three to six students each to the senior class. Among the army students, officers from branches besides the coast artillery corps such as cavalry, field artillery, air corps, and corps of engineers attended the senior course. The army did not send officers to the junior class. The army service school for captains and majors was the Command and General Staff College in Fort Leavenworth, Kansas. Only one coast guard officer attended the War College during the middle phase—one lieutenant junior grade was a member of the junior class of 1933. He would be the last representative from his service to attend for the remainder of the interwar period. Table 5 documents the demographics of the middle phase War College junior and senior classes.

The first year of the late phase was also the first year of the War College advanced course. The Taussig Board first recommended a separate course for flag officers and senior captains in its 1929 report, but the advanced course never did achieve the vision of the progression of War College presidents who were its advocates.[4] Advanced course students were generally already senior course graduates, and they spent their year conducting independent study and did not participate in wargames. One thing the advanced course did accomplish was to serve as a host for all flag officer students of the remainder of the interwar period, and admirals or generals no longer attended the senior course. Senior class size remained high, reaching an interwar period high of fifty-four in 1938, and line commanders continued to make up the bulk of the class. Army and marine corps student attendance increased slightly, from six to eight students of each service per class. Junior class size greatly decreased during this phase, due to the need for junior officers to fill billets aboard new ships built as part of the Vinson-Trammel Act of 1934. Overall class size and makeup remained consistent until the class of 1941, when the entire student body was reduced to approximately half of the average class size up to that time. Table 6 illustrates the demographics of the late phase War College classes, including the advanced classes.

Table 4. Early phase student body composition

	Jun 1919	Dec 1919	Jun 1920	1921	1922	1923	1924 SR	1924 JR	1925 SR	1925 JR	1926 SR	1926 JR	1927 SR	1927 JR
U.S. Navy														
RADM		2		1			1		2		1		1	
CAPT	22	12	15	11	8	13	14		16		10		15	
CDR	5	10	8	9	25	24	28		23		26		17	
LCDR					5	5		11		10	2	16		16
LT								9		7		6		8
LTJG										1				
U.S. Marine Corps														
BGen														
Col			1		2	2	3		1		2		1	

LtCol	1		1	1	2	1						1		2	
Maj	1		2	1			2		4				3		
Capt															
U.S. Coast Guard															
All Ranks			1	1	1			1							
U.S. Army															
COL	1		2	1	3		2								
LTC				1	1	3	1			2			1		
MAJ	1		2	2	1	1	1			2			1	1	
CPT														1	

A table listing the composition of the Naval War College student body during the early phase, from 1919 to 1927. *Source:* Naval War College, *Register of Officers, 1884–1955.* JR = junior course, SR = senior course.

Table 5. Middle phase student body composition

	1928 SR	1928 JR	1929 SR	1929 JR	1930 SR	1930 JR	1931 SR	1931 JR	1932 SR	1932 JR	1933 SR	1933 JR	1934 SR	1934 JR
U.S. Navy														
RADM					1		1		1					
CAPT	15		16		8		9		8		12		7	
CDR	19		26		31		23		22		25		34	
LCDR		22		17		13		10		6	1	7	1	12
LT		14		35		22		29	1	30		32		23
LTJG												1		
U.S. Marine Corps														
BGen														
Col	2		1		1				1		1			

	1	2	3	4	5	6	7	8
LtCol		1	2		2	1	2	
Maj	1	1			1	2	2	
Capt						1	1	1
U.S. Coast Guard								
U.S. Army								
COL				1				
LTC	1	2	2		2		1	
MAJ	2	2	3		4	6	5	
CPT								

A table listing the composition of the Naval War College student body during the middle phase, from 1928 to 1934. *Source:* Naval War College, *Register of Officers, 1884–1955.* JR = junior course, SR = senior course.

Table 6. Late phase student body composition

	1935 AD	1935 SR	1935 JR	1936 AD	1936 SR	1936 JR	1937 AD	1937 SR	1937 JR	1938 AD
U.S. Navy										
RADM	1			1			1			
CAPT	4	5		2	9		9	8		4
CDR	4	25		4	25			30		
LCDR		1	17			5		1	4	
LT			23			17			6	
LTJG										
U.S. Marine Corps										
BGen				1			1			
Col	1			1	1			1		1
LtCol		2		1	3			2		
Maj		2			1			2		
Capt		1	1							
U.S. Coast Guard										
U.S. Army										
COL	1			1						
LTC					2		1	3		1
MAJ		5			3			3		
CPT		1			3					

A table listing the composition of the Naval War College student body during the late phase, from 1935 to 1941. *Source*: Naval War College, *Register of Officers, 1884–1955*. AD = advanced course, JR = junior course, SR = senior course.

1938 SR	1938 JR	1939 AD	1939 SR	1939 JR	1940 AD	1940 SR	1940 JR	1941 AD	1941 SR	1941 JR
					2					
12		6	6		3	7		2	8	
27		1	34			28			10	
2	4			2			6			5
	11			15			11			
1										
		1			1	2		2	2	
2			4			3			3	
4			3			3			2	
4		1	3		1	3		1	1	
2			3			2			2	
						1				

Appendix B

Naval War College Wargames

The table below is an attempt to consolidate all the operations, strategic, and tactical maneuvers held at the Naval War College between 1919 and 1941 into one listing. It builds on a similar table included as an appendix to Michael Vlahos's *The Blue Sword: The Naval War College and the American Mission, 1919–1941*, with additional information extracted from records maintained in Record Group 4, Publications, 1914–74, housed at the Naval Historical Center at the Naval War College. All the information in the table was extracted from maneuver summaries, critiques, and blueprint histories of maneuvers.

Maneuver problems of the interwar period were designated as either strategic (sometimes abbreviated as STRAT or S), or tactical (abbreviated as TAC or T) until 1927, when the "operations" (abbreviated OP) designation took the place of "strategic." For a short period, some operations games encompassed first a strategic chart maneuver and then a tactical board maneuver based on the end state of the chart maneuver, though occasionally War College classes played tactical board maneuvers in the absence of a strategic predecessor. The problems issued to each class were numbered serially with Roman numerals according to the class period in which they were used. Until 1930 certain problems that were used from year to year were also given permanent Arabic identification numbers followed by modification numbers in case the same problem was used in a modified form.

Where appropriate and where discovered in the course of research for this book, predecessor maneuvers, as well as salient notes regarding maneuver problems, are listed. Demonstration, search and screening, and quick-decision games are not included in this listing, and the games that are listed are not necessarily in chronological order.

Abbreviations

AB: advance base
AWC: Army War College
BB: battleship
CA: cruiser
CC: battle cruiser
CLV: aircraft-carrying cruiser
CP N: convoy problem north
CV: aircraft carrier
CVE: escort carrier (converted merchant ship)
FAM: familiarization
FLT: fleet
IO: Indian Ocean
IVO: in vicinity of
JR: junior
LANT: Atlantic Ocean
MB: main body
NM: nautical miles
PAC: Pacific Ocean
PI: Philippine Islands
SA: South America
SE: southeast
SLOC: sea line of communication
WESTPAC: western Pacific Ocean

Table 7. Wargames at the Naval War College, 1919–41

Class of 1919-1						
Type	Number		Predecessor	Opponent	Notes	
S	40	V		ORANGE		
S	49	VII		ORANGE		
S	56	VIII		ORANGE		
T	2	I		RED		
T	2	III		RED		
T	2	IV		RED		
T	7	V		RED		
T	7	VI		RED		
T	21	VII		RED		
T	49	IX		RED		
T	53	VIII		RED		
T	74	XI		ORANGE		
T	75	XI		ORANGE		
T	76	XII		RED		
T	77	XIII		RED		

Class of 1919-2

Type	Number		Predecessor	Opponent	Notes
S	36	V		ORANGE	Hawaii-Panama
S	40	VI		ORANGE	
S	43	I		ORANGE	
S	49	IV		ORANGE	
S	57	IX		ORANGE	
S	58	VII		ORANGE	West PAC
T	10	I		RED	
T	13	III		RED	
T	14	II		RED	Red superior skill
T	26	V		RED	
T	27	VI		RED	
T	71	VII		RED	
T	74	X		ORANGE	
T	79	XI		RED	
T	81	XIV		ORANGE	

Class of 1920-1

Type	Number	Predecessor	Opponent	Notes
S	35	III	ORANGE	
S	40	VI	ORANGE	
S	43	I	ORANGE	Hawaii-Panama
S	49	IV	ORANGE	
S	56	VIII	ORANGE	
S	57	IX	ORANGE	
T	10	I	RED	
T	12	IV	ORANGE	
T	26	V	RED	
T	27	VI	RED	
T	49	IX	RED	
T	77	XIII	RED	
T	78	VII	RED	
T	81	X	ORANGE	
T	82	XV	RED	

Class of 1920-2

Type	Number		Predecessor	Opponent		Notes
S	35	III		RED		
S	43	I		ORANGE		Hawaii-Panama
S	44	II		RED		
S	62	V		RED		
S	63	VI		RED		
S	49	IV		ORANGE		
T	10	I		RED		
T	14	II		RED		Red superior skill
T	26	IV		RED		
T	77	VII		RED		
T	78	VI		RED		

Class of 1921

Type	Number		Predecessor	Opponent		Notes
S	35	VII		ORANGE		
S	64			ORANGE		
T	27	V		RED		

Type	Number	Predecessor	Opponent	Notes
T	49	VIII	RED	
T	83		ORANGE	

Class of 1922

Type	Number	Predecessor	Opponent	Notes
S	44	II	ORANGE	
S	49	IV	ORANGE	Train OP
S	57	VII	ORANGE	
S	65	III	ORANGE	Panama
S	66	V	ORANGE	
S	67	VI	ORANGE	PAC north route
S	68	V	ORANGE	
T	84	III	ORANGE	FLT standing order
T	85	IV	ORANGE	
T	86	V	ORANGE	
T	87	VI	RED	Based on Jutland
T	88	VII	ORANGE	
T	89	VIII	ORANGE	Treaty navy
T	90	IX	ORANGE	
T	91	X	ORANGE	

Class of 1923

Type	Number		Predecessor	Opponent	Notes
S	70	A		BLACK	Azores
S	70	B		ORANGE	
S	71	I		ORANGE	
S	72	II		ORANGE	
S	74	III		ORANGE	Chart Maneuver II
S	75	IV		ORANGE	Formosa
S	76	V-A		ORANGE	
S	76	V-B		ORANGE	Logistics
T	92	II		RED	
T	93	III		RED	
T	94	IV		RED	Battle of Emerald Bank
T	96	V		ORANGE	Battle of Marianas

Class of 1924

Type	Number		Predecessor	Opponent	Notes
S	72	II		ORANGE	Search OP
S	74	IV		ORANGE	

Type	Number		Predecessor	Opponent	Notes
S	75	V		ORANGE	North route to Avacha Bay
S	77	B		ORANGE	
S	79	VI		ORANGE	14-day prep from Hawaii
T	10 mod 8	I		RED	
T	10 mod 9	II		RED-CRIMSON	Battle of Sable Island
T	79	IV		ORANGE	
T	96	III		ORANGE	Battle of Siargao

Class of 1925

Type	Number		Predecessor	Opponent	Notes
S	72 mod 2	II		ORANGE	Japan home waters
S	74 mod 1	V		ORANGE	PI, antilanding
S	76 mod 2	IV		ORANGE	
S	77	B		ORANGE	
S	80			ORANGE	Joint Problem 1, PI, antilanding
T	96	III		ORANGE	
T	98	II		RED	Caribbean
T	101			ORANGE	PI

Class of 1926

Type	Number		Predecessor		Opponent	Notes
OP	1.2	I			ORANGE	
S	69 mod 2	A			BLUE	
S	72 mod 3	II			ORANGE	
S	74.2				ORANGE	
S	77	B			ORANGE	
T	10.8	I			RED	
T	101	III			ORANGE	Convoy attack
T	102	IV			ORANGE	

Class of 1927

Type	Number		Predecessor		Opponent	Notes
OP		I	Search and screening	5.9	ORANGE	
OP		II			ORANGE	
OP		III			BLACK	
S	72.4	I			ORANGE	Convoy Protection
S	77.3	B			ORANGE	BLUE advance across PAC
T	104	I			ORANGE	
		C			ORANGE	Air raid on Pearl Harbor

Class of 1928

Type	Number		Predecessor	Opponent	Notes
OP	1	I		ORANGE	Relief expedition convoy defense
OP	2	II		ORANGE	Truk–Malampaya
OP	3	III		BLACK	RED forced landing IVO New Bedford
OP	4	IV		ORANGE	Middle phase of BLUE versus ORANGE
OP	5	V		ORANGE	Paper exercise—committee study
T	104.1	I		ORANGE	

Class of 1929

Type	Number		Predecessor	Opponent	Notes
OP	2	II		ORANGE	
OP	3	III		ORANGE	BLUE SLOC ORANGE bases flanking
OP	4	IV		ORANGE	Joint army-navy; amphibious operations with CVE support
OP	5	V		ORANGE	Initial phase: bases for advance
OP	6	VI	OP IV-28	ORANGE	Joint w/AWC; forced landing
S		II		ORANGE	
S		IV		ORANGE	

S		v		ORANGE	
S	7	VII		RED	
T	2	II		ORANGE	
T	5	V		RED	
T	104.1	I		ORANGE	
T	4	IV		ORANGE	
T		VI		ORANGE	
T		VII		ORANGE	

Class of 1930

Type	Number	Predecessor	Opponent	Notes
OP		I	ORANGE	Asiatic FLT defense of insular territory
OP		II	RED	BLUE scouting force versus RED expeditionary force off CRIMSON
OP	3	III	ORANGE	ORANGE north, BLUE south; w/JR class after 1 month play
OP	4	IV	RED	Panama
OP		VI	ORANGE	Batangos landing, support group
STRAT		V	ORANGE	Paper exercise
TAC		I	ORANGE	Truk

Type	Number		Predecessor		Opponent	Notes
TAC	2	A	OP	III	ORANGE	
TAC	3	B	OP	IV	RED	
TAC	4	C	OP	VI	ORANGE	

Class of 1931

Type	Number	Predecessor	Opponent	Notes
OP	II		ORANGE	BLUE 3xCLV and 1xCV
OP	III		RED	RED AB at Trinidad, similar to OP IV-30
OP	IV		ORANGE	
TAC	I		ORANGE	

Class of 1932

Type	Number	Predecessor	Opponent	Notes
OP	I		ORANGE	Asiatic FLT
OP	II		ORANGE	Suez and IO transit to PAC
OP	III		RED	South LANT/IO, plan change in midcourse
OP	IV		ORANGE	
OP	V		ORANGE	Revision of OP IV movement
TAC	I		ORANGE	ORANGE offensive, BLUE defensive
TAC	II		ORANGE	

Type	Number	Predecessor	Opponent	Notes
TAC	III		ORANGE	BLUE superiority
TAC	IV	OP III	RED	Trinidad-Liberia
TAC	V		RED	Battle of Sable Island
TAC	VI	OP V	ORANGE	

Class of 1933

Type	Number	Predecessor	Opponent	Notes
OP	I		ORANGE	Refueling at sea
OP	II		ORANGE	Same as OP II-32; Suez to South PI
OP	III		RED	Azores
OP	IV		ORANGE	Last thruster strategy maneuver
TAC	I		ORANGE	Same as TAC I-32
TAC	II		ORANGE	Same as TAC II-32, Truk
TAC	III	OP III	RED	
TAC	IV		RED	Battle of Sable Island
TAC	V	OP IV	ORANGE	

Class of 1934

Type	Number	Predecessor	Opponent	Notes
OP	I		ORANGE	

Type	Number	Predecessor		Opponent	Notes
OP	II			ORANGE	Asiatic FLT OP Plan
OP	III	OP	IV-33	ORANGE	Continuation of OP IV-33 SR
OP	III			ORANGE	
OP	IV			RED	2nd phase, east LANT
OP	V			ORANGE	1 year after hostilities, PURPLE ally
TAC	I			ORANGE	
TAC	II			ORANGE	Truk
TAC	III	OP	III	ORANGE	
TAC	IV	OP	IV	RED	
TAC	V			RED	Battle of Sable Island
TAC	VI	OP	V	ORANGE	Java Sea and Kamchatka, PURPLE ally

Class of 1935

Type	Number	Predecessor		Opponent	Notes
OP	I			ORANGE	Begins after AB established at Dumanquillas
OP	II			RED	Caribbean area defense; CRIMSON neutral
OP	III			ORANGE	Trans PAC strategic area FAM/FLT composition
STRAT	I			ORANGE	Intro to estimate and orders

Type	Number	Predecessor		Opponent	Notes
STRAT	II			ORANGE	Asiatic FLT problem
TAC	I	STRAT	I	ORANGE	Manila
TAC	II			ORANGE	Truk area; at war for 6 months
TAC	III	OP	II	RED	Halifax
TAC	IV	OP	III	ORANGE	Flows from OP III, BLUE MB transit from Truk to PI

Class of 1936

Type	Number	Predecessor		Opponent	Notes
OP	I			ORANGE	Strategic areas IVO BROWN; 1 month operations by BLUE
OP	II			ORANGE	AB cut ORANGE SLOCs east and south China Sea
OP	III	OPS	I, II	ORANGE	Battle IVO Pellew as BLUE MB nears PI
OP	IV	OPS	I, II, III	ORANGE	BLUE MB at Dumanquilas; 3 convoys arrive from US via Suez, ORANGE to intercept
STRAT	II			RED	
TAC	I			ORANGE	CC action San Bernadino Strait; escort force for BLUE convoy to Manila
TAC	II			ORANGE	Truk

Type	Number	Predecessor	Opponent	Notes
TAC	III		RED	FLT battle
TAC	IV		ORANGE	BLUE FLT divided, ORANGE Trans PAC offense before BLUE can reunite near Canal Zone

Class of 1937

Type	Number	Predecessor	Opponent	Notes
OP	I	OP	ORANGE	Same as 1936 OP, BLUE CA raiding ORANGE SLOCs in South China Sea
OP	II		ORANGE	Same as OP I, advance of BLUE MB on Truk
OP	III		ORANGE	BLUE MB from Truk to PI; PURPLE involvement
OP	IV		ORANGE	Air and surface patrols; Dumanquillas-Suez
STRAT	I		ORANGE	Same as 1936 OP
STRAT	II		RED	BLUE raiding of SLOCs weakens FLT, RED assumes offensive
TAC	II		WHITE	Hypothetical opponent, like ORANGE but larger
TAC	III		ORANGE	CP N of Truk/ORANGE force Truk to Dumanquilas
TAC	IV		RED	Red MB on offensive to keep BLUE from taking Halifax

Class of 1938

Type	Number	Predecessor	Opponent	Notes
OP	I		ORANGE	Aleutians

Type	Number	Predecessor	Opponent	Notes
OP	II		RED	Defensive campaign in Caribbean
OP	III		ORANGE	OP plan
OP	IV		ORANGE	Area FAM safeguarding route
OP	V		ORANGE	Capture/defense of ABs
OP	VI		RED	Sable Island, RED BB over 60% damage
OP	VII		ORANGE	SILVER and BLACK aiding ORANGE, BROWN ally
OP	VIII		ORANGE	BLUE covering AB Staring Bay, ORANGE seeking decisive action

Class of 1939

Type	Number	Predecessor	Opponent	Notes
OP	I		ORANGE	BLUE base in Aleutians
OP	II		RED	Caribbean
OP	III		ORANGE	Raid on ORANGE SLOCs in IO, China Sea
OP	IV		ORANGE	Truk
OP	V		ORANGE	ORANGE MB at Saipan, BLUE at Pearl
OP	VI		RED	North LANT area FAM
OP	VII		ORANGE	Have been at war 2 years, BLUE-BROWN secret alliance

Class of 1940

Type	Number	Predecessor	Opponent	Notes
OP	I		ORANGE	Aleutians
OP	II		RED	Caribbean
OP	III		ORANGE	Raid on ORANGE SLOCs in West PAC and China Sea
OP	IV		ORANGE	Truk

Class of 1941

Type	Number	Predecessor	Opponent	Notes
OP	I		ORANGE	Aleutians
OP	II		BLACK-SILVER	BLACK in North LANT, Narvik, Iceland, SILVER-GOLD in Med to sortie to SA, GOLD bases in Caribbean, monopoly of SA trade
OP	III		ORANGE	Minus new LANT FLT, ORANGE threaten Aleutian-Wake-Samoa Line, BLUE to take Eniwetok
OP	IV		ORANGE	Defense of Truk, commitments in LANT/Brazil prevent BLUE offensive, BLUE losses high, ORANGE low, ORANGE attack w/ BBs versus 2 BLUE BB
OP	V		ORANGE	WESTPAC, China Sea, raid on ORANGE SLOC, BLUE fully committed in LANT, unable to force decision in PAC, RED neutral, BLUE has Truk to hold line to Australia, 4 CA 1 CV from Truk to Java Sea
OP	VI		BLACK-SILVER	BLACK holds EMERALD; RED ally

Class of 1941

Type	Number	Predecessor	Opponent	Notes
OP	II		BLACK-SILVER-GOLD	BLACK trade monopoly of SA; RED ally
OP	II		ORANGE	Truk defense against ORANGE, BLUE 2 *North Carolina* class BB
OP	III		ORANGE	Raid on ORANGE SLOCs, rendezvous 6/19 Coral Sea, 7/150 NM SE Singapore
OP	IV		BLACK	BLUE and RED versus BLACK North LANT 40-ship convoy

A table listing the strategic and tactical wargames maneuvered at the Naval War College during the interwar period, 1919 through 1941. The table lists the game type, its numerical designation, the BLUE opponent, and notes regarding the game scenario. Source: Vlahos, *Blue Sword*, 198, appended with additional information extracted from records maintained in RG4, Publications, 1914–74, NHC.

Appendix C

The World Naval Balance, 1919–41

Student-officers attending the Naval War College during the interwar period conducted wargames that pitted BLUE forces split between two oceans against numerically superior RED forces in the Atlantic and more modern ORANGE forces in the Pacific. These game conditions reflected perceptions of the real-world naval balance by maneuver staff and students. This balance, or lack thereof, and its ultimately destabilizing effect, was a result of limited budgets, an aging fleet, internationally mandated limits to naval strength, and a number of other factors outside the typical naval officer's sphere of influence. However, the students were well aware of them, and their perceptions influenced their strategy and tactics in games, as well as their comments in postgame critiques and memoirs. This appendix summarizes the evolving naval balance during the interwar period and describes how the wargames reflected this balance.

When the First World War ended in 1918, Germany and Austria-Hungary were finished as naval powers. On the winning side, however, the story was not so clear. With a force of forty-four capital ships (battleships and battle cruisers) and the world's first aircraft carriers, Great Britain held sway as the world's preeminent naval power, but the war had drained it both materially and economically. France, with seven capital ships in its navy, was in much the same condition. Italy, with five capital ships, was a distant third among the European allies in terms of naval strength.[1] The United States, entering the war in 1917, had avoided the large manpower losses experienced by its allies, and its industrial base was untouched by the physical effects of war. American industrial mobilization, the Naval Act of 1916 (which resulted in the laying down of ten dreadnought battleships that year alone), and the expansionist sea power doctrines of Alfred Thayer Mahan placed the U.S. Navy in a commanding position once the Armistice was signed on 11 November 1918. Because of this surge in shipbuilding, by 1919 the U.S. Navy boasted a force of seventeen capital ships, most of which featured modern oil-fired engines.

The new variable in the balance of naval power was Japan. Since its victory over Russia in the Russo-Japanese War of 1904–5, it had steadily built up its Pacific fleet, which was able to play a major role in the region during World War I. Shortly after its declaration of war against Germany in late 1914, Japanese forces occupied German territories on the islands of Palau and Yap; in the Marianas, Marshall, and Caroline island chains; and at the port city of Rabaul on the island of New Britain (figure 29). Japan entered into an informal agreement with Great Britain that Commonwealth forces would not claim any of the numerous German Pacific island possessions north of the equator.[2] The League of Nations formally recognized Japan's occupation of these islands under a Class C Mandate. In the 1930s Japan fortified many of these islands as naval and air bases. It was through these island bases that the BLUE fleet of the interwar period wargames had to pass to reach the Philippines. Foremost among them was Truk in the Caroline Islands. With its large natural harbor, Truk became a forward base for the Japanese fleet and figured prominently in ORANGE wargames. Guam, the only U.S. territory in this area, lay to the northwest, where it was surrounded by Japanese-controlled island bases.

In response to British requests for material assistance during the war, and in exchange for cooperation in territorial moves both in the central Pacific and on the Shantung peninsula in China, Japan deployed a naval task force consisting of a cruiser and eight destroyers to bolster Royal Navy operations in the Mediterranean theater. By the end of the war this force had grown to seventeen Japanese ships (plus some British vessels manned by Japanese crews) that escorted 788 transports carrying seven hundred thou-

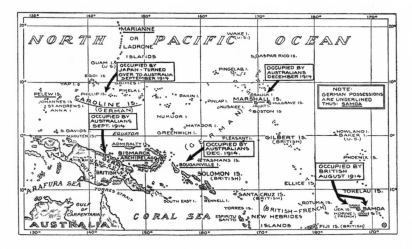

FIG. 29. A contemporary map showing German colonies in the Pacific during the First World War and how the Allies dealt with them. "The European War," *New York Times Current History* 2, no. 5 (August 1915).

sand British Commonwealth troops. Japanese naval vessels also escorted convoys transporting troops from Australia through the Indian Ocean and patrolled as far as South Africa and Mexico.[3]

By 1919 the Imperial Japanese Navy was the world's fourth largest, with nine capital ships. In 1922 the new battleships *Nagato* and *Mutsu*, with sixteen-inch guns and flank speeds of almost twenty-eight knots, augmented this already substantial naval force. The western Allies initially appreciated and encouraged Japan's developing military strength, but eventually they came to see it as a threat to their own Pacific colonial possessions such as British Hong Kong, U.S. territory in the Philippines, French Indochina, and the Dutch East Indies.[4]

After the end of the war the governments of most of the combatant nations viewed a large military as an unnecessary and potentially destabilizing expense and demobilized large portions of their armed forces. During a sharp depression in 1921, these same governments also questioned the need for expensive ongoing and future shipbuilding programs. Great Britain especially opposed a naval arms race with the United States, which she was then economically unable to match. While the United States was able to outproduce its naval rivals, the Harding administration, elected on a platform of a "return to normalcy," had little inclination to do so. Traditional American aversion to large standing armed forces had an immediate effect on both the army and the navy in the years following the Armistice, as those services commenced drastic reductions in force. Furthermore, General William "Billy" Mitchell and other aviation advocates trumpeted the potential of air power as a less expensive and more effective alternative to battleships for defending U.S. shores from hostile forces. Japan, while in a new position of strength, had one-third of its economy tied up in naval construction and could not maintain such a pace indefinitely. Too much force displayed too soon by the Japanese could also prompt Great Britain and France—now that they did not have to maintain large forces in European waters to counter Germany—to redeploy their navies to the Pacific if they believed their overseas interests were threatened. Accordingly, Japan was also in a mood to negotiate at least a slowdown in naval construction.[5]

In 1921 Britain was ready to propose a naval arms limitation conference among the leading post–World War I naval powers. The intent was to develop an agreement to slow the burdensome arms race and reduce the possibilities for future wars. In the United States, under pressure from the Senate, the Harding administration beat the British to the punch and offered to host such a conference in Washington. The original intention was for the conference to address all armaments, but the focus narrowed quickly to naval forces. Nine nations were invited—the five naval powers of Britain, France, Italy, Japan, and the United States, plus China, Portugal, Belgium, and the Netherlands. The former Central Powers and the Soviet Union were excluded from the conference.

Several separate agreements were signed at the conclusion of the Washington Conference. The naval treaty, known as the Five Power Treaty on Naval Limitation, was signed on 6 February 1922. All signatories pledged to maintain a balance in their respective capital fleets under a specified ratio. The treaty ratios were 5 for Britain and the United States, 3 for Japan, and 1.67

each for France and Italy. The ratios of Britain and the United States were the largest because of their need to maintain large navies in both the Atlantic and the Pacific Oceans. All signatories agreed to honor a capital ship construction "holiday" for ten years.[6] Japan had negotiated hard for a 3.5 ratio (equivalent to 70 percent of that of the U.S. or Royal Navies), but in the end was compelled to accept a tonnage ratio of 60 percent.[7] To help Japan overcome its reluctance to accept this limit, the major Pacific naval powers—Britain, Japan, and the United States—agreed not to increase fortifications on most of their Pacific bases.

The Washington Treaty was significant in what it did *not* restrict. Attendees discussed restricting submarine warfare but never came to an agreement. Aircraft carriers were ostensibly limited under Article VII of the treaty, but none of the signatories had anywhere near the tonnage limits prescribed (in 1922 there were only five aircraft carriers in the entire world). Signatories could convert two battle cruisers currently under construction to aircraft carriers, but even with these additions, there was still sufficient tonnage remaining for each nation to build new ones. These terms did not greatly concern naval planners of 1922, who viewed the aircraft carrier as an experimental, defensive weapon of little value in a major naval war. But most significant was that the Washington Treaty contained no enforcement mechanism or system of checks and balances, relying instead on the word of each signatory that they would honor the treaty terms as written.

In accordance with the terms of the new agreement, all the major maritime powers voluntarily reduced the size of their navies. The British scrapped or stopped ongoing construction on twenty-four ships, and the Japanese did the same on sixteen.[8] For the U.S. Navy the impact of this agreement was immediate and dramatic. Sixteen existing battleships were decommissioned (three predreadnoughts were expended during the Billy Mitchell tests in 1921 and 1923), and construction of six new battleships and four new battle cruisers that had been started between 1920 and 1921 was halted.[9]

The Royal Navy held on to its position as the world's largest in terms of both tonnage and numbers during the 1920s, but these figures are misleading, as a large percentage of this fleet dated from the First World War. One modern battle cruiser (*Hood*) and two new battleships (*Nelson* and *Rodney*) were placed in service during this period, but the rest of the capital ship fleet received bare-minimum modernization such as conversion from coal to oil-fired engines. Like the United States and Japan, Great Britain converted capital ship hulls to aircraft carriers (*Courageous* and *Glorious*), but the results were far inferior in performance and equipment to those of the other countries. The other European signatories of the Five Power treaty, France and Italy, had their own economic and political difficulties during the twenties and did not embark on major naval expansion efforts until the thirties, though France did convert one battleship into an aircraft carrier (*Bearn*).

In contrast with the western powers, the Japanese Navy expanded to the maximum extent of its treaty limitations. Like the United States and Great Britain, Japan converted capital ship hulls, which otherwise would have been scrapped, into aircraft carriers (*Akagi* and *Kaga*) and embarked on an ambitious program to build new cruisers (a ship class unregulated by the 1922 treaty) and destroyers. Between 1923 and 1928 Japan added eleven heavy and four light cruisers to its naval strength.[10] At the time of their launching these ships were the most advanced of their class in the world. Naval historian Anthony Preston noted that "Japanese warship design took a direction all its own, in pursuit of a doctrine which demanded absolute superiority in each category. Thus each cruiser class had to be superior to any foreign cruiser, and each destroyer had to match any foreign destroyer, with no great regard for what was needed. . . . No navy has ever approached its problems in such a dogmatic and doctrinaire fashion."[11] Beyond mere shipbuilding, Japan aggressively modernized its naval equipment, training, and tactics, developing lethal long-range torpedoes, becoming proficient at night operations, and holding large combined fleet exercises. These improvements would pay major dividends in combat during the first years of World War II.

This renewed escalation did not entirely escape notice. American and British naval intelligence experts were aware that Japan was improving its military capabilities, and they made occasional naval appropriations—most notably for construction of cruisers—to counter this increase. But the western powers were hampered by an inability or unwillingness to take the Japanese threat seriously. Anthony Preston's comments on the American intelligence assessment of the new *Yubari* class cruiser were typical: "The designed standard tonnage of the . . . *Yubari* was 2890 tons. This figure was the one which was published, whereas the ship actually displaced 17 percent more. Puzzled Western naval intelligence departments wrestled with the *Yubari*'s staggering figures and dreamed up ludicrous reasons for the marked discrepancy between Japanese and Western designs of a similar size, the most popular being that the small stature of Japanese sailors permitted the designers to squeeze the internal dimensions."[12] Because of this surge in the construction of cruisers by the Washington Treaty signatories, President Calvin Coolidge felt compelled to convene another naval summit in Geneva in 1927. This conference was not successful, as France and Italy refused to attend and the countries that did attend, particularly Great Britain and Japan, were unable to agree on tonnage ratios.[13] As a result, cruiser construction remained unregulated.

While the U.S. Navy modernized its battleship fleet, by failing to build up to the treaty limits, it fell far behind the other treaty signatories in new ship construction. BLUE fleet composition in wargames reflected this disparity. Game planners occasionally augmented BLUE fleets with additional ship types such as the CLV to allow students to experiment with new concepts, but generally, they stayed true to real force levels. With the inauguration of Franklin D. Roosevelt and the initiation of New Deal industrial programs, the American shipbuilding industry would receive a massive shot in the arm that eventually propelled the U.S. Navy back into a position of naval supremacy. Nevertheless, in 1933 the United States had a lot of ground to make up, and it would be nine years before its navy would be capable of sustaining the trans-Pacific advance envisioned in War Plan ORANGE.

From 1922 until 1933 the Royal Navy remained superior in

numerical terms but slipped notably in terms of quality. For example, Britain possessed the greatest number of aircraft carriers (six) of all the major powers, but the aircraft they embarked were much fewer and less capable than their Japanese or American counterparts. Great Britain continued to place great faith in the battleship as the centerpiece of the Royal Navy and had more on hand than any other combatant nation when World War II came in September 1939, though all but two of these ships were of World War I vintage. They maintained a numerical advantage in battleships over the U.S. Navy until the 1930s (figure 30), and this difference, coupled with the U.S. policy of maintaining most of their battle force in the Pacific, led to RED versus BLUE wargames that generally turned out in favor of RED.

While Japan complied with the letter of the 1922 treaty, its ambitious approach to military expansion, coupled with its willingness to circumvent or ignore specific treaty conditions, gave that country a distinct advantage in terms of individual ship quality. American intelligence officers knew of the differences in technical quality and reflected it in ORANGE fleets. Aspects the games did not capture were that Japan equipped and trained its naval personnel to a higher degree of readiness than either the United States or Great Britain did, and its aggressive moves in China provided both strategic and tactical experience in real-world combat operations that its future opponents lacked. By 1933 the Imperial Japanese Navy had the strength sufficient to allow Japan to challenge the western powers openly.

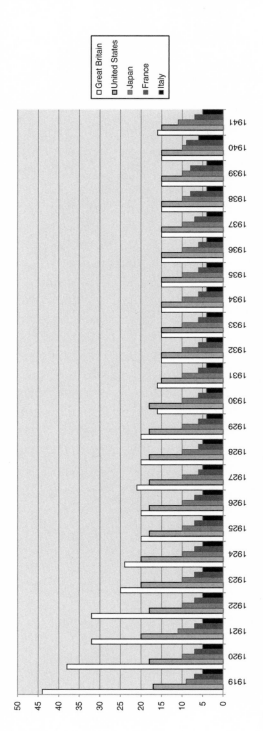

FIG. 30. A chart showing the relative numbers of battleships and battle cruisers belonging to the five nations that signed the Washington Naval Treaty during the years between the world wars. Breyer, *Battleships and Battle Cruisers*.

Notes

Abbreviations

LOC   Library of Congress, Washington DC
NHC   Naval Historical Collection, Naval War College, Newport RI
NHHC   Naval History and Heritage Center, Washington DC

Introduction

1. Rear Admiral Eller's quote is from Hoyt, *How They Won the War*, viii.

2. Potter, *Nimitz*, 136; Buell, interview by Stilwell, 247; and Smith, "Preparing for War," 1.

3. Mahan, *Influence of Sea Power*.

4. To trace the development of the American vision for a Pacific war, see Miller, *War Plan ORANGE*.

5. Expanded discussion of these and other innovations in naval warfare are investigated in Murray and Millet, *Military Innovation*; Kuehn, *Agents of Innovation*; and Hone and Hone, *Battle Line*.

6. See Perla, *Art of Wargaming*, 15–34, for a thorough summary of the European origins of wargaming.

7. Von Reisswitz, *Anleitung zur Darstellung*.

8. The definition of the War College mission is from Stephen Bleeker Luce to Boutelle Noyeu, 19 July 1885, in Gleaves, *Life and Letters*, 158. The characterization of the War College is from Hattendorf, Simpson, and Wadleigh, *Sailors and Scholars*, xiiv.

9. For an in-depth discussion of nineteenth-century wargaming in the United States and its adoption at the Naval War College, see Spector, *Professors of War*.

10. Dring, interview by Nicolosi, 13.

11. U.S. Naval War College, "Chronology of Courses and Significant Events, 1890–1899," U.S. Naval War College website, accessed 13 October 2012, https://www.usnwc.edu/About/History/Chronology-of-Courses-and -Significant-Events/1890s.aspx.

12. McCarty Little to Admiral Stephen Luce, 9 August 1895, Reel 9, Box 10, Stephen Bleeker Luce Papers, Manuscript Division, Naval Historical Foundation Collection, LOC.

13. Theodore Roosevelt to Caspar Goodrich, Theodore Roosevelt Papers, Series 2, 16 June 1897, Reel 313, Manuscript Division, LOC.

14. See Lea, *Day of the Saxon*, and Bywater, *Great Pacific War*.

15. Spector, *Professors of War*, 88–111; U.S. Naval War College, "NWC History," U.S. Naval War College website, accessed 23 August, 2010, http://www.usnwc.edu/About/History.aspx.

16. Cline, *Washington Command Post*, 29.

17. John Hattendorf compiled statistics regarding the number of Naval War College students who achieved flag rank for Dr. Evelyn Cherpak, head of the NHC.

18. Richardson, *On the Treadmill*, 109.

19. Secretary of the Navy, memorandum for the president, 9 March 1942, President's Secretary's File, Safe File, Navy Department: March–September 1942, Box 4, Franklin D. Roosevelt Presidential Library, described in Frank, "Picking Winners?," 24–30.

20. Spector, *Professors of War*, 144–55.

21. Vlahos, *Blue Sword*, iv, 31.

22. Vlahos, *Blue Sword*, 105–10.

23. In a 1929 speech to the War College, chief of naval operations Admiral W. V. Pratt stated, "For these two services [the U.S. and Royal Navies] to engage in war would be little short of criminal . . . throwing the entire world into chaos out of which nothing but world revolution would come." From Pratt, "The Aspects of Higher Command," address delivered before the U.S. Naval War College, 30 August 1929, Folder 1, Box 17, William V. Pratt Papers, NHHC, 11–12.

24. Vlahos, "Wargaming," 7–22.

25. Halsey and Bryan, *Admiral Halsey's Story*, 54.

26. Schifferle, *America's School for War*, 31–34.

27. Felker, *Testing American Sea Power*, 3.

28. Nofi, *To Train the Fleet for War*, 20–28.

29. Nofi, *To Train the Fleet for War*, 45n39.

30. Miller, *War Plan ORANGE*, xix–xxi.

31. Miller, *War Plan ORANGE*, 17.

32. Kuehn, *Agents of Innovation*, xiv.

33. Hone and Hone, *Battle Line*, 183.

34. Perla, *Art of Wargaming*, 8.

35. Vlahos, "War Gaming," 7–22.

36. Signatories of the Five Power Treaty on Naval Limitation pledged to maintain a balance in their respective capital fleets under specified ratios of five for Britain and the United States, three for Japan, and 1.67 each for France and Italy. All signatories also agreed to honor a ten-year capital ship construc-

tion "holiday." See "Five Power Treaty on Naval Limitation, 6 February 1922," *Statutes at Large of the United States*, 43, Pt. 2, 1655–85.

37. The Vinson-Trammel Act, otherwise known as the Naval Parity Act, was signed into law on 27 March 1934. It authorized the construction of 1,184 naval aircraft and 102 warships (sixty-five destroyers, six cruisers, thirty submarines, and one aircraft carrier) to be started over the next three years and completed by 1942. Between May and December 1940, both Holland and France had surrendered to Germany, Japan had deployed troops and aircraft into French Indochina, and the Tripartite Pact had been signed. In November 1940, chief of naval operations Admiral Harold Stark determined that War Plan ORANGE was no longer executable, and he submitted proposals for a new global strategy (originally named War Plan DOG, then later RAINBOW) that called for a holding action in the Pacific until the situation in Europe was stabilized.

38. The NHC contains more than twelve hundred linear feet of records documenting the administrative and curricular history of the institution since its founding in 1884. The archives house forty-five record groups, including administrative correspondence, curriculum items and publications, conference proceedings, library records, lectures, faculty and staff presentations, theses, World War II Battle Evaluation Group records, and intelligence and technical source materials pertaining to technological developments and strategic and tactical problems of interest to the navy. See Cherpak, *Guide to Archives*, 3, 7, 216.

39. Harris Laning, "Information and Instructions for Student Officers, June 1923" Folder 925, Box 21, RG4, Publications, 1915–77, NHC, 12.

1. The Players

1. Mahan, "Necessity and Object," 626.

2. William S. Sims, "The Practical Character of the Naval War College," lecture at the U.S. Naval Academy, Annapolis MD, November 11, 1912, U.S. Naval War College website, accessed 26 November 2011, http://www.usnwc .edu/getattachment/About/History/SimsDoc.pdf.aspx; Richardson, *On the Treadmill*, 109.

3. Sims, "Practical Character of the Naval War College," 1.

4. Todd, *Study and Discussion*, 2.

5. U.S. Naval War College, *Estimate of the Situation*.

6. U.S. Naval War College, *The Mission and Organization of the Naval War College, 1936–1937*, Folder 108, Box 7, Collection 619, Papers of Dewitt C. Ramsey, 1914–49, Operational Archives Branch, NHHC, 11.

7. Both dotter and subcaliber exercises were simulations devised to help naval gun crews practice full-scale gunnery operations. Admiral Percy Scott RN developed dotter exercises for the Royal Navy. He described them as follows: "On a vertical board, opposite to the muzzle of the gun, was a metal frame which . . . could be moved up and down at either a slow or a fast rate. On

this frame was painted a bull's-eye, and beside it was a card with a line drawn upon it. On the face of the board, and moved either up or down by the muzzle of the gun, was a carrier containing a pencil. When the men under instruction pressed the trigger of the gun the pencil, actuated by an electrical contrivance, made a dot on the card, and ... moved a space to the right. If the gun was truly pointed at the bull's-eye at the moment of firing, the dot would be in line with the bull's-eye. If the gun was not truly pointed, the amount of error was indicated on the card." From Admiral Sir Percy Scott, *Fifty Years in the Royal Navy* (London: John Murray, 1919), 87. Subcaliber exercises involved the firing of a smaller-caliber round from a special shell casing in a large-caliber gun.

8. Sims "Practical Character of the Naval War College," 12–14.

9. W. S. Sims to secretary of the navy, 15 January 1919, with endorsement from W. S. Benson, 23 January 1919, in *Outline History of the Naval War College, 1884 to 1937*, unpublished compilation, Naval War College Library, Newport RI, 88–99.

10. U.S. Naval War College, *Mission and Organization of the Naval War College*, 1.

11. Pratt biographer Gerald Wheeler wrote, "Once drafted, almost all the war plans were tried on the game board at the War College.... Pratt definitely helped establish the War College as a place where the Department's war plans could be legitimized." From Wheeler, *Admiral William Veazie Pratt*, 244.

12. Edward Kalbfus to J. W. Greenslade, 15 July 1935, Folder 1, Box 4, John Wills Greenslade Papers, 1847–1961, Manuscript Division, LOC.

13. Gaining "familiarity with the composition of existing fleets" and "actual strategical areas" are listed as one of the "motives" behind maneuvers in U.S. Naval War College, *Mission and Organization of the Naval War College*, 5.

14. Miller, *War Plan ORANGE*, 17.

15. Thomas H. Robbins Jr. to Thomas Buell, 20 October 1970, Folder 15, Box 3 MS Coll. 37, Thomas Buell Papers, NHC.

16. U.S. Naval War College, *Register of Officers, 1884–1977*. Joseph Taussig commanded the first American destroyer squadron to deploy to England, and Reginald Belknap was commander of the force that laid the North Sea mine barrage, the U.S. Navy's only major operation of the war. See Still, *Queenstown Patrol, 1917*, and Belknap, *Yankee Mining Squadron*.

17. Hattendorf, Simpson, and Wadleigh, *Sailors and Scholars*, 142–46, and Nofi, *To Train the Fleet for War*, 38.

18. James L. Holloway to Thomas B. Buell, 7 October 1970, Folder 13, Box 1, MS Coll. 37, Thomas Buell Papers, NHC.

19. U.S. Naval War College, "Hints on the Solution of Tactical Problems," by Harris Laning, Folder 788, Box 18, RG4, Publications, 1915–77, NHC.

20. Wildenberg, *All the Factors of Victory*, 256–58.

21. Withers, "Preparation of the Submarines Pacific for War," 387–93, and Weir and Allard, *Building American Submarines*, 40–41.

22. Miller, *War Plan ORANGE*, 137.

23. W. V. Pratt, "The Aspects of Higher Command," address delivered before the U.S. Naval War College, 30 August 1929, Folder 1, Box 17, William V. Pratt Papers, NHHC, 11–12.

24. Wheeler, *Admiral William Veazie Pratt*, 243.

25. Raymond A. Spruance to president, Naval War College, 3 November 1965, Folder 7, Box 1, MS Coll. 12, Raymond Spruance Papers, NHC.

26. Holloway to Buell, 7 October 1970.

27. Moore, interview by Mason, 537.

28. Bieri, interview by Mason, 78; Bernhard H. Bieri to George Dyer, "Recollections of R. K. Turner," 22 May 1976, Folder 45, Box 1, MS Coll. 37, Thomas Buell Papers, NHC; Moore, interview by Mason, 536–37.

29. "List of Books on History of War," Folder 1078-E, Box 25, RG4, Publications, 1915–77, NHC.

30. U.S. Naval War College, "The Battle of Marianas as Maneuvered at the U.S. Naval War College by the Class of 1923: History and Tactical Critique," by Harris Laning, Folder 779, Box 17, RG4, Publications, 1915–77, NHC, 96; Captain R. B. Coffey, Tactical Problem VI-1932 (Op. Prob. V), "Critique," Folder 1718-E, Box 58, RG4, Publications, 1915–77, NHC, 38; Tactical Problem III-1936-SR, "Critique," Folder 2021-F, Box 74, RG4, Publications, 1915–77, NHC, 3.

31. Educational Department, U.S. Naval War College, "General Description of the Course in Lectures, Theses, and International Law," June 1934, Folder 1870, Box 67, RG4, Publications, 1915–77, NHC.

32. Rear Admiral William Moffett, chief of the Bureau of Aeronautics and the navy's senior ranking aviator, presented the aviation lecture on 9 February 1923. His opposite number in submarines, Captain George Day, presented the submarine lecture on 16 February. On 23 March the class received the lecture titled "Late Developments in Armor" from Lieutenant Commander David I. Hedrick of the Bureau of Ordnance. See U.S. Naval War College, "Program for March 1923, Class of 1923," by D. W. Blamer, Folder 857, Box 19, RG4, Publications, 1915–77, NHC.

33. J. Q. Dealey, "Survey of the World Situation as It Affects the United States," lecture presented at the Naval War College 5 August 1927, Folder 1261, Box 33, RG4, Publications, 1915–77, NHC. Professor Huntington gave the geography/sociology lecture on 4 August 1922. See U.S. Naval War College, "Program for August 1922, Class of 1923."

34. See Karsten, *Naval Aristocracy*, for a complete demographic analysis of the interwar naval officer corps.

35. In addition to their navy and marine corps officers who made up the majority of classes, each Naval War College class contained a certain number of officers from the army and, until 1925, one or two officers of the U.S. Coast Guard. See appendix A for interwar period class demographics.

36. Downs, *Naval Personnel Organization*, 52–53.

37. The annual report of the Secretary of War for 1927, the year after the last "emergency (World War non-regular) officer" was discharged from service, listed total officer strength at 11,913, of whom 3,394 received their commissions through the Military Academy at West Point NY. U.S. War Department, *Report of the Secretary of War*, 187.

38. Future World War II naval leaders who sailed with the Great White Fleet included Ensign Harold R. Stark and Midshipman Raymond A. Spruance on USS *Minnesota* (BB-22) and Midshipman William F. Halsey on USS *Kansas* (BB-21).

39. Thomas Hart would go on to command the battleship *Mississippi* (BB-41), and Nimitz commanded the heavy cruiser *Augusta* (CA-31).

40. Dyer, *Amphibians Came to Conquer*, 124.

41. The naval aviation observer course was established in 1922 to provide aviation training for nonpilot flight crew members on multiplace aircraft. The observer course also provided a necessary qualification to senior officers who lacked the aviation designation required by law to make them eligible to command the new aircraft carriers.

42. Bieri, interview, 81.

43. Potter, *Nimitz*, 125.

44. Kimmel, *Admiral Kimmel's Story*, 3.

45. Buell, *The Quiet Warrior*, 24, 42.

46. One of the most frequent users of historical references was Captain R. B. Coffey '22, who made multiple references to Horatio Nelson (TAC VI-32) and Civil War land battles (TAC IV-33) to emphasize points in his postgame critiques.

47. Richardson, *On the Treadmill to Pearl Harbor*, 109–10; Hustvedt, interview, 182; and Moore, interview by Mason, 520–21.

48. U.S. Naval War College, *Mission and Organization of the Naval War College*, 1, and Bieri, interview, 76.

49. Laning, "Information and Instructions for Student Officers," 9.

50. Dring, interview by Nicolosi, 14; Gaudet, interview by McHugh, 31.

51. Laning, "Information and Instructions for Student Officers," 1.

52. U.S. Naval War College, "Extracts from Books Read in Connection with War College Reading Courses," vols. 1 and 2, by H. R. Stark, Folder 880, Box 20, RG4, Publications, 1915–77, NHC.

53. Vlahos, *Blue Sword*, 75–85; Hattendorf, Simpson, and Wadleigh, *Sailors and Scholars*, 125–27, 142.

54. Floor plan of Luce Hall and Mahan Hall, Folder 1270, Box 33, RG4, Publications, 1915–77, NHC.

55. Dring, interview, 12–14; Gaudet, interview, 6.

56. U.S. Naval War College, *Conduct of Maneuver*, 1928 edition, Folder 1399, Box 44, RG4, Publications, 1915–77, NHC, 18.

57. U.S. Naval War College, "The Chart and Board Maneuvers," June 1928, Folder 1398, Box 44, RG4, Publications, 1915–77, NHC, 41.

58. Martin, interview by McHugh, 11–12.

59. Moore, interview by Mason, 536; Gaudet, interview, 12.

60. Dring, interview, 3–4.

61. Laning, "Information and Instructions for Student Officers," 2–10.

62. Moore, interview, 534–35.

63. Entry of 31 July 1924–17 November 1925, personal diary of Harry L. Pence, Series 4, Harry Pence Papers, 1893–1976, MSS 0144, Mandeville Special Collections Library, Geisel Library, University of California, San Diego.

64. Entry of 17 March 1923, diary of Thomas Hart, Folder January 1, 1923–December 30, 1923, Box 8, Papers of Admiral Thomas C. Hart, 1899–1960, Operational Archives Branch, NHHC.

65. Health-related entries in the Hart diary for the winter of 1922–23 include family member bouts with colds, influenza, whooping cough, and bronchitis, as well as a mumps scare. Hart also wrote of the death of the infant daughter of Reginald Belknap from an unspecified illness on 28 December. Mrs. Nimitz's discussion of life in Newport is contained in Mrs. Catherine Nimitz to Stansfield Turner, 2 April 1973, Folder 1, Box 1, RG29, Students, Staff & Faculty, NHC.

66. Dring, interview, 2.

67. Hart diary, 8 August and 21 November 1922, and 2 February 1923, NHHC.

68. U.S. Naval War College, "Program for March 1923, Class of 1923."

69. U.S. Naval War College, "Introductory Remarks, Tactics Course," by R. C. MacFall, Folder 1898, Box 67, RG4, Publications, 1915–77, NHC, 2.

70. U.S. Naval War College, "Introductory Remarks, Tactics Course," 5; Department of Operations, U.S. Naval War College, "Operations Problem V-1932-SR; Notes Taken at Critique," Folder 1722H, Box 59, RG4, Publications, 1915–77, NHC, 15.

71. Moore, interview, 521.

72. Endorsements of this viewpoint from William Halsey '33 and Ernest King '33 are contained in Halsey and Bryan, *Admiral Halsey's Story*, 54, and King and Whitehill, *Fleet Admiral King*, 242.

2. The Game Process

1. U.S. Naval War College, *The Mission and Organization of the Naval War College, 1936–1937*, Folder 108, Box 7, Collection 619, Papers of Dewitt C. Ramsey, 1914–49, Operational Archives Branch, NHHC, 6.

2. John A. Battilega and Judith K. Grange, "The Misappropriation of Models," in Hughes, *Military Modeling*, 281–90.

3. Perla, *Art of Wargaming*, 205–14.

4. Game board re-creations of World War I naval battles figured prominently in the class schedule, but they were played according to a script to illustrate how a real-world situation would look on the game board. These "demonstration maneuvers" did not involve any decision making on the part of the players. See Gaudet, interview by McHugh, 9–10, 19.

5. In 1903 the Joint Board started the convention of using different colors to designate war plans against different countries. Late in the interwar period, "Rainbow" was a code name for a multination war.

6. Vlahos, *Blue Sword*, 103, 107.

7. William Veazie Pratt, "The Aspects of Higher Command," address delivered before the U.S. Naval War College, 30 August 1929, Folder 1, Box 17, William V. Pratt Papers, NHHC, 11–12.

8. U.S. Naval War College, "Operations Problem IV-1933-SR; Critique by Captain Todd," by Forde A. Todd, Folder 2261AA, Box 93, RG4, Publications, 1915-77, NHC, 1.

9. The ORANGE scenarios made extensive use of real-world research and intelligence. For example, the operations staff maintained a detailed survey of potential anchorages throughout the entire Pacific area to assist them in selecting bases for scenarios. U.S. Naval War College, "Possible Insular Anchorages in the Pacific Ocean, 17 May 1920," Folder 1360-9, Box 37, RG4, Publications, 1915-77, NHC.

10. Vlahos, "Wargaming," 11.

11. This is an important difference between interwar games and modern games, where the opposition is played by "aggressors" or a "Red Cell" of players who are usually staff members and who consciously play their roles as they predict that the real-world enemy would.

12. Buell, "Admiral Raymond A. Spruance," 47–48.

13. U.S. Naval War College, "The Conduct of Maneuvers (Introductory)," October 1928 edition, Folder 1399, Box 44, RG4, Publications, 1915-77, NHC, 20.

14. For a complete listing of the strategic, tactical, and operations games of the interwar period, refer to Appendix B.

15. U.S. Naval War College, "Quick Decisions," by J. W. Wilcox, Folder 1321, Box 35, RG4, Publications, 1915-77, NHC, and U.S. Naval War College, "General Procedure of the Conduct of 'Quick Decision' Problems," by S. C. Rowan, Folder 1723, Box 59, RG4, Publications, 1915-77, NHC, 2–3.

16. Operations problems were combinations of strategic and tactical maneuvers. See chapter 4.

17. U.S. Naval War College, "Conduct of Maneuvers (Introductory)," 2, 4.

18. U.S. Naval War College, *Mission and Organization of the Naval War College*, 3–5; Bieri, interview by Mason, 78.

19. U.S. Naval War College, "Maneuver Rules, Tactics Jacket; June 1928," Folder 1439-4, Box 46, RG4, Publications, 1915-77, NHC.

20. William S. Sims, "The Practical Character of the Naval War College," lecture at the U.S. Naval Academy, Annapolis MD, 11 November 1912, U.S. Naval War College website, accessed 26 November 2011, http://www.usnwc .edu/getattachment/About/History/SimsDoc.pdf.aspx, 3–8; Harris Laning, "The Naval Battle," appendix to *Admiral's Yarn*, 409–10.

21. Laning, "Naval Battle," 273.

22. U.S. Naval War College, "General Information; War Game Outfit, Mark III, December 1930," Folder 6, Box 1, RG35, Naval War Gaming, 1916–2003, NHC.

23. U.S. Naval War College, *Mission and Organization of the Naval War College*, 6.

24. Martin, interview by McHugh, 5.

25. U.S. Naval War College, "Conduct of Maneuvers (Introductory)," 25.

26. Dring, interview by Nicolosi, 17.

27. U.S. Naval War College, "Conduct of Maneuvers (Introductory)," 23; Gaudet, interview, 3–4.

28. Figure 8 was taken with a fisheye lens, which makes the maneuver room appear larger than it actually was. It also shows two sixteen-by-twenty-five-foot maneuver boards laid side by side. The inscription on the photograph reads "Game board not of sufficient size to permit the playing of such a tactical problem as that of the Battle of Jutland."

29. U.S. Naval War College, "Conduct of Maneuvers (Introductory)," 27–28.

30. Navy Department Bureau of Yards and Docks, Drawing 114764, "Naval War College Newport RI, Extension of Main Building; Second Floor Plan," Folder 9, Drawer 5, Folio III, NHC.

31. U.S. Naval War College, "The Conduct of Maneuvers," 1934 edition, Folder 1879, Box 67, RG4, Publications, 1915–77, NHC, 26.

32. U.S. Naval War College, "Conduct of Maneuvers," 26; Public Works Drawing 2528-35, "Naval War College, Newport RI Alterations to Main Building, Plan of Third Floor," Folder 10, Drawer 5, Folio File III, NHC.

33. U.S. Naval War College, "Conduct of Maneuvers (Introductory)," 7, 12–13.

34. Dring interview, 13; U.S. Naval War College, "Conduct of Maneuvers (Introductory)," 32–38.

35. U.S. Naval War College, "Conduct of Maneuvers (Introductory)," 28.

36. Captain Harry L. Pence recorded that "a well-balanced maneuver will require roughly one-third of the class for umpiring duty. . . . No particular duty in a maneuver should be repeated by any officer during the College course." From U.S. Naval War College, "Suggestions for Conducting Chart and Board Maneuvers," by Harry L. Pence, Folder 1461, Box 48, RG4, Publications, 1915–77, NHC, 3.

37. Vlahos, *Blue Sword*, 137.

38. U.S. Naval War College, "Conduct of Maneuvers (Introductory)," 13; and U.S. Naval War College, "Maneuver Rules," Rule E-8, Folder 1439-4, Box 46, RG4, Publications, 1915–77, NHC, 67–68.

39. Gaudet, interview, 12.

40. Martin, interview, 4.

41. Gaudet, interview, 4.

42. U.S. Naval War College, "Conduct of Maneuvers (Introductory)," 51, 52; Gaudet, interview, 20.

43. U.S. Naval War College, "Conduct of Maneuvers (Introductory)," 38.

44. Gaudet, interview, 6; U.S. Naval War College, "Conduct of Maneuvers (Introductory)," 15, 20.

45. Dring, interview, 17.

46. Moore, interview by Mason, 535-36.

47. U.S. Naval War College, "Conduct of Maneuvers (Introductory)," 8.

48. Entry of 28 August 1922, diary of Thomas Hart, Folder January 1, 1922–December 30, 1922, Box 8, Papers of Admiral Thomas C. Hart, 1899–1960, Operational Archives Branch, NHHC.

49. Hart diary, September 1922, NHC.

3. The Early Phase

1. U.S. Marines were deployed to intervene in small-scale conflicts so often during this period that they developed a formal doctrine for such operations. Major S. M. Harrington's 1921 report "The Strategy and Tactics of Small Wars" would evolve into an official *Small Wars Manual* by 1940.

2. President, General Board, to Secretary of the Navy, 30 July, 12 October, and 9 November 1915, General Board File 420-2, RG 80.7.3, Records of the General Board, 1900-1951, National Archives and Records Administration, College Park MD.

3. Between 1915 and 1918, the U.S. Navy increased in size from 32 battleships to 39 and from 57 destroyers to 110. From Naval History and Heritage Command, "U.S. Navy Active Ship Force Levels, 1886–present," Naval History and Heritage Command website, accessed 26 October 2012, http://www.history.navy.mil/branches/org9-4.htm#1910.

4. McHenry, *Webster's American Military Biographies*, 392.

5. Five Power Treaty on Naval Limitation, 6 February 1922, Chapter I, Article II, *United States Statutes at Large*, 43, Pt. 2, pp. 1655-85.

6. Miller, *War Plan ORANGE*, 77-80.

7. U.S. Naval War College, "Outline of the Course in Tactics; Lecture Presented to the Class of 1924," by Harris Laning, Folder 800, Box 18, RG 4, Publications, 1915-77, NHC.

8. Wheeler, *Admiral William Veazie Pratt*, 67.

9. Ellis, *Advanced Base Operations in Micronesia*.

10. Kuehn, *Agents of Innovation*, 130-43.

11. In Fleet Problem I, a single plane launched from the battleship USS *Oklahoma*, representing an entire carrier air group, dropped ten practice bombs and theoretically "destroyed" the spillway of the Gatun Dam. See MacDonald, *Evolution of Aircraft Carriers*, 30.

12. During OP IV 28, BLUE forces featured not only 152 destroyers at a time when active strength of this type was only 103 but also a fleet train of 198 ships when the total number of naval auxiliaries of all types was just 68.

13. Hattendorf, Simpson, and Wadleigh, *Sailor and Scholars*, 131–32.

14. Borneman, *Admirals*, 131, and Potter, *Nimitz*, 138.

15. U.S. Naval War College, "Tactical Problem II (Tac. 92); Maneuvered 23–29 September 1922 as Tactical Maneuver II," Folder 830, Box 19, RG4, Publications, 1915–77, NHC.

16. U.S. Naval War College, "Strategic Problem III (Strat .74); Maneuvered 20 October–11 November 1922 as Chart Maneuver II," Folder 820, Box 18, RG4, Publications, 1915–77, NHC.

17. Potter, *Nimitz*, 138.

18. U.S. Naval War College, "The Battle of Emerald Bank as Maneuvered at the U.S. Naval War College by the Class of 1923: History and Tactical Critique," by Harris Laning, Folder 774, Box 17, RG4, Publications, 1915–77, NHC, 2.

19. Harris Laning, "The Naval Battle," appendix to *Admiral's Yarn*, 423–27.

20. U.S. Naval War College, Memorandum 1100 (f) 11–22, "Class of 1923 Tactical Problem IV (Tac. 94), Subject: Tactical Problem IV," Folder 774, Box 17, RG4, Publications, 1915–77, NHC, 12; U.S. Naval War College, "Battle of Emerald Bank as Maneuvered at the U.S. Naval War College by the Class of 1923," 78–80.

21. U.S. Naval War College, Memorandum 1100 (f) 11–22, "Class of 1923 Tactical Problem IV (Tac. 94), Subject: Tactical Problem IV."

22. U.S. Naval War College, "Battle of Emerald Bank, Tactical Problem IV, TAC 94, 1923, Diagram 9," Folder 832, Box 19, RG4, Publications, 1915–77, NHC.

23. U.S. Naval War College, "Battle of Emerald Bank as Maneuvered at the U.S. Naval War College by the Class of 1923," 11–12.

24. Ibid., 4.

25. U.S. Naval War College, "Battle of Emerald Bank, Tactical Problem IV, TAC 94, 1923, Diagram 36."

26. Entry of 9 December 1922, diary of Thomas Hart, Folder January 1, 1922–December 30, 1922, Box 8, Papers of Admiral Thomas C. Hart, 1899–1960, Operational Archives Branch, NHHC; U.S. Naval War College, memorandum 1100 (f) 11–22, "Class of 1923 Tactical Problem IV (Tac. 94), Subject: Tactical Problem IV," 12; U.S. Naval War College, "Battle of Emerald Bank as Maneuvered at the U.S. Naval War College by the Class of 1923," 78–80.

27. Thomas C. Hart diary, 13 January 1923, NHHC.

28. U.S. Naval War College, "Battle of Emerald Bank as Maneuvered at the U.S. Naval War College by the Class of 1923," 18.

29. U.S. Naval War College, "Battle of Emerald Bank," 80.

30. U.S. Naval War College, "The Battle of Marianas as Maneuvered at the U.S. Naval War College by the Class of 1923: History and Tactical Critique," by Harris Laning, Folder 779, Box 17, RG4, Publications, 1915–77, NHC, 1. "Japanese Mandate Islands" refers to former German Pacific Territories of the Marianas, Carolines, Marshall Islands, and Palau groups held by Japan

under a class C mandate in accordance with article 22 of the covenant of the League of Nations. See appendix C.

31. In his diary entries of 29 January and 26 February 1923, Thomas Hart mentions that he is playing the role of Japanese commander in chief.

32. U.S. Naval War College, "Battle of Marianas as Maneuvered at the U.S. Naval War College by the Class of 1923," 3.

33. Compare Hornfischer, *Neptune's Inferno*, 299–310, with U.S. Naval War College, "The Battle of the Marianas, Class of 1923, U.S. Naval War College, Tactical Problem V (Tac. 96), 23 May 1923," by Harris Laning, Folder 779, Box 17, RG4, Publications, 1915–77, NHHC, 84–94.

34. Thomas C. Hart diary, 27 March 1923, NHHC.

35. U.S. Naval War College, "Battle of Marianas as Maneuvered at the U.S. Naval War College by the Class of 1923."

36. Nofi, *To Train the Fleet for War*, 28. Laning would serve two years as Tactics Department head and would return to the War College in 1931 as president.

37. U.S. Naval War College, "Battle of Marianas as maneuvered at the U.S. Naval War College by the Class of 1923," 95.

38. "Army and Navy War Game," *Time Magazine*, Monday, 30 May 1927; "Defenders of Narragansett Bay Engaged in Mighty Effort to Locate Enemy Fleet off the New England Coast," *Indiana Evening Gazette*, 17 May 1927; and "Ship Is Sunk in Play War: Army-Navy Maneuvers are Held Along East Coast of New England," *Sarasota Herald Tribune*, 19 May 1927, 2.

39. U.S. Naval War College, "Class of 1927 Operations Problem No. III; The Study of Overseas Expedition with Forced Landing; General Situation and Special Situation BLACK," Folder 1289, Box 35, RG4, Publications, 1915–77, NHC, 2–4.

40. U.S. Naval War College, "Class of 1927 Operations Problem No. III," 7–8.

41. U.S. Naval War College, "Battle of Marianas as Maneuvered at the U.S. Naval War College by the Class of 1923," 96.

42. U.S. Naval War College, "Battle of Emerald Bank as Maneuvered at the U.S. Naval War College by the Class of 1923," 18.

43. U.S. Naval War College, "Battle of Marianas as Maneuvered at the U.S. Naval War College by the Class of 1923," 94.

4. The Middle Phase

1. Colton, "U.S. Navy Ships and Submarines," Shipbuilding History: Construction Records of U.S. and Canadian Shipbuilders and Boatbuilders, accessed June 10, 2012, http://www.shipbuildinghistory.com/history/navalships.htm.

2. Naval History and Heritage Command, "U.S. Navy Active Ship Force Levels, 1886-present," Naval History and Heritage Command website, accessed 27 March 2008, http://www.history.navy.mil/branches/org9-4.htm.

3. The *Congressional Record* of 19 July 1930 contains the following passage from a speech by Senator David I. Walsh (D-MA): "Following that conference [Washington] and up to January 1, 1929, the great Powers of the world laid down and appropriated for naval expansion as follows: Japan, 125 naval vessels: Great Britain, 74 naval vessels: France, 119: Italy, 82: and, to the everlasting credit of our own country, the United States, exclusive of small river gunboats, 11."

4. U.S. Naval War College, *Register of Officers, 1884–1955*, NHC.

5. Captain Stephen C. Rowan paraphrased in U.S. Naval War College, "Operations Problem II-1933-SR, Comment of the Research Department," Folder 1779G, Box 61, RG4, Publications, 1915–77, NHC, 5; U.S. Naval War College, "Operations Problem IV-1928-SR, History of Maneuver," section (i) Conclusions and Lessons Learned, Folder 1382-V1, Box 41, RG4, Publications, 1915–77, NHC, 6.

6. For the class of 1927, Strategic Problem I (Strat. 72) took forty-five hours over twelve calendar days (equating to 155 three-minute moves) to play seven hours and forty-five minutes of "game time." See Research Department, U.S. Naval War College, "Analysis of Trans-Pacific Problems as Played at Naval War College, Newport," Folder 2261-1q, Box 94, RG4, Publications, 1915–77, NHC. Improvements in game time and other areas are discussed in U.S. Naval War College, "Operations Problem IV-1928-SR, History of Maneuver," section (i), Conclusions and Lessons Learned, 4–6.

7. U.S. Naval War College, "Class of 1928 Operations Problem No. III; The Study of Overseas Expedition with Forced Landing," Folder 1381, Box 39, RG4, Publications, 1915–77, NHC, 1; U.S. Naval War College, "Operations Problem IV-1928-SR," Conclusions and Lessons Learned, 2.

8. BLUE approach routes along the Aleutian chain to the Kamchatka Peninsula were first evaluated by the class of 1922 in S.67/STRAT VI, and finally by the class of 1934 in TAC VI. In a probable reference to S.75/STRAT IV, Thomas Hart '23 recorded in his diary, "Finished my 'problem' today. A plan for getting at the Japs via Aleutian Islands, Kamchatka, etc. I've been very much interested and now think that though not good it's perhaps our best chance of doing something if we ever have that war." Entry of 22 December 1922, diary of Thomas Hart, Box 8, Folder January 1 1922–December 30, 1922, Papers of Admiral Thomas C. Hart, 1899–1960, Operational Archives Branch, NHHC.

9. Vlahos, "Wargaming," 11–13.

10. U.S. Naval War College, "The Chart and Board Maneuvers," June 1928 edition, Folder 1398, Box 44, RG4, Publications, 1915–77, NHC, 44.

11. Wildenberg, *Gray Steel and Black Oil*, 42.

12. U.S. Naval War College, "Blueprint Record of Moves, OP IV 1928; 25 March 1933, Numerical Comparison and General Situation," Folder 1382-V2, Box 41 RG4, Publications, 1915–77, NHC.

13. Department of the Navy Bureau of Yards and Docks, *Building the Navy's Bases in WWII*, vol. 1, *History of the Bureau of Yards and Docks, and the*

Civil Engineering Corps, 1940–1946 (Washington DC: Government Printing Office, 1947), 209.

14. Air-capable ships in the wargame fleets appear under a number of designations, including some that are not obvious. First-line aircraft carriers were designated "CV," second line were "OCV," and auxiliary carriers were "XOCV." In OP IV 29, the designation AV referred to a "Heavier-Than-Air Tender." After 1937 the AV designation was assigned specifically to seaplane tenders. From U.S. Naval War College, "Conduct of Maneuvers," 1928 edition, Folder 1399, Box 44, RG4, Publications, NHC, 25.

15. U.S. Naval War College, "Operations Problem III-1932; Critique," by Forde A. Todd, Folder 1721-I, Box 57, RG4, Publications, 1915-77, NHC, 58.

16. U.S. Naval War College, "Naval War College Course, 1927–28," by R. Z. Johnston, Folder 1252, Box 33, RG4, Publications, 1915-77, NHC, 2.

17. U.S. Naval War College, "Outline of War College Course 1927–1928, and Details under Div. 'C' of Senior Class and Joint Junior-Senior Class Work," Folder 1330, Box 36, RG4, Publications, 1915-77, NHC.

18. U.S. Naval War College, "Class of 1928 Operation Problem No. III; The Study of Overseas Expedition with Forced Landing."

19. U.S. Naval War College, "Operations Problem IV-1928-SR (Trans-Pacific problem)," Folder 1382-VI, Box 41, RG 4, Publications, 1915-77, NHC, 4.

20. Ibid.

21. U.S. Naval War College, "Class of 1928 (Senior) Division 'C' Movement, (Sections 1 and 2), Operations problem IV, Period 17 January–4 February 1928," by J. W. Greenslade, Folder 1382-V1, Box 41, RG4, Publications, 1915-77, NHC.

22. U.S. Naval War College, "Operations Problem IV-1928-SR (Trans-Pacific problem)," 7 of Section (i) Conclusions and Lessons Learned.

23. U.S. Naval War College, "Blueprint Record of Moves, OP IV 1928; Moves 38–40, 25 March 0815 to 25 March 0830," Folder 1382-V2, Box 41, RG4, Publications, 1915-77, NHC.

24. U.S. Naval War College, "U.S. Army War College—U.S. Naval War College Operations Problem VI 1929; Joint Army and Navy Operations with Forced Landing; Part 1, The Preliminary Situation, 1 April," Folder 1438A, Box 46, RG4, Publications, 1915-77, NHC, 1. OP IV 28 was the predecessor game to OP VI 29; this source was prepared before the class of 1929 convened to orient that class to their scenario.

25. U.S. Naval War College, "Operations Problem IV-1928-SR (Trans-Pacific problem)," 2–5, 11 of Section (i) Conclusions and Lessons Learned.

26. Ibid., 8–9 of Section (i) Conclusions and Lessons Learned.

27. Ibid., 15 of Section (i) Conclusions and Lessons Learned.

28. Ibid., 18 of Section (i) Conclusions and Lessons Learned.

29. Blair, *Silent Victory*, 27.

30. Research Department, U.S. Naval War College, "Analysis of Trans-Pacific Problems as played at Naval War College, Newport," Folder 2261-1q, Box 94, RG4, Publications, 1915-77, NHC.

31. U.S. Naval War College, "Operations Problem II–1931–SR; History and Critique of the Tactical Phase," by Benjamin Dutton, Folder 1647-I, Box 54, RG4, Publications, 1915–77, NHC.

32. U.S. Naval War College, "Tactical Problem I-1932-SR," by R. B. Coffey, Folder 1709-F, Box 57, RG4, Publications, 1915–77, NHC.

33. "Navy Pioneer Is Dead—Capt. Wilbur R. Van Auken—40 Years in Service, Was 71," *New York Times*, 15 August 1953.

34. Laning, *Admiral's Yarn*, 334.

35. The statement "It is believed that the reduction of the RED air superiority is more important than the reduction of the RED DD superiority" is another indication of the increasing awareness of the utility of carrier-based aviation. U.S. Naval War College, "Tactical Problem IV-1932-SR & JR; Critique," by R. B. Coffey, Folder 1714-G, Box 57, RG4, Publications, 1915–77, NHC, 5.

36. U.S. Naval War College, "Operations Problem II–1931–SR; History and Critique of the Tactical Phase."

37. Ibid.

38. U.S. Naval War College, "Tactical Problem II-1932-SR; Critique," by R. B. Coffey, Folder 1710-I, Box 57, RG4, Publications, 1915–77, NHC.

39. U.S. Naval War College, "Senior Class of 1932, Operations Problem II, History and Critique of Chart Maneuver," Folder 1720-E, Box 57, RG4, Publications, 1915–77, NHC, 38–39.

40. U.S. Naval War College, "Tactical Problem II-1932-SR; Critique," Section VI, "BLUE and ORANGE Submarines," and U.S. Naval War College, "Operations Problem V-1932-SR; Notes Taken at Critique," Folder 1720-E, Box 57, RG4, Publications, 1915–77, NHC, 9.

41. Captain Joseph Taussig to Captain John Greenslade, 26 February and 10 October 1929, Box 1, John Wills Greenslade Papers, Naval Historical Foundation Collection, Manuscript Division, LOC.

42. U.S. Naval War College, "Operations Problem III-1932; Critique," by Forde A. Todd, Folder 1721-I, Box 57, RG4, Publications, 1915–77, NHC, 13.

43. U.S. Naval War College, "Operations Problem II-1933-SR; Comment of the Research Department," by Wilbur Van Auken, Folder 1779G, Box 61, RG4, Publications, 1915–77, NHC, 4.

44. U.S. Naval War College, "Operations Problem II-1933-SR," 3.

45. U.S. Naval War College, "Operations Problem II-1933-SR," 3, 8–9.

46. U.S. Naval War College, "Tactical Problem IV-1933-SR; Critique," by R. B. Coffey, Folder 1791E, Box 62, RG4, Publications, 1915–77, NHC.

47. U.S. Naval War College, "Operations Problem IV-1933-SR; Critique by Captain Todd," by Forde A. Todd, Folder 2261AA, Box 93, RG4, Publications, 1915–77, NHC, 1.

48. Hattendorf, Simpson, and Wadleigh, *Sailors and Scholars*, 144–45.

49. U.S. Naval War College, "Operations Problem IV-1933-SR; Critique by Captain Todd," 1–3.

50. U.S. Naval War College, "Operations Problem IV-1933; Summarized Data," Folder 2261-1T, Box 94, RG4, Publications, 1915–77, NHC, Enclosure "T."

51. U.S. Naval War College, "Operations Problem IV-1933-SR; Stenographic Notes Taken at Critique," Folder 2261-1n, Box 94, RG4, Publications, 1915–77, NHC, 8–9. Major Bradley was junior in rank but senior in years to his classmates, being a graduate of the Naval Academy class of 1910. Captain Whiting was naval aviator number 16. See "Major General Follett Bradley," U.S. Air Force website, Biographies, accessed 16 August 2015, http://www.af.mil/AboutUs/Biographies/Display/tabid/225/Article/108025/major-general-follett-bradley.aspx; "Kenneth Whiting," *Dictionary of American Naval Fighting Ships*, accessed 14 September 2015, http://www.history.navy.mil/research/histories/ship-histories/danfs/k/kenneth-whiting.html.

52. U.S. Naval War College, "Operations Problem IV-1933-SR; Stenographic Notes Taken at Critique," Folder 2261-1n, Box 94, RG4, Publications, 1915–77, NHC, 10–11.

53. U.S. Naval War College, "Operations Problem IV-1933-SR/Tactical Problem V-1933-SR; Comment upon Battleships, Air Forces, Cruisers, and Train of BLUE—General Remarks upon Other Phases in Operations Problem IV and Tactical Problem V," by Wilbur R. Van Auken, Folder 2261-10, Box 94, RG4, Publications, 1915–77, NHC, 8–9.

54. Research Department, U.S. Naval War College, "Analysis of Trans-Pacific Problems as played at Naval War College, Newport," Folder 2261-1q, Box 94, RG4, Publications, 1915–77, NHC.

55. Doyle, "The U.S. Navy and War Plan Orange," 55.

56. Research Department, U.S. Naval War College, "Senior Class of 1934; Operations Problem III and Tactical Problem III; Analysis and Comment of the Research Department," Folder 1858-E, Box 65, RG4, Publications, 1915–77, NHC.

5. The Late Phase

1. The Naval Expansion Act of 1934, HR 6604, 27 March 1934, otherwise known as the First Vinson-Trammel Act, authorized President Roosevelt "to undertake . . . the construction of: (a) One aircraft carrier of approximately fifteen thousand tons standard displacement, to replace the experimental aircraft carrier Langley; (b) ninety-nine thousand two hundred tons aggregate of destroyers to replace over-age destroyers; (c) thirty-five thousand five hundred and thirty tons aggregate of submarines to replace over-age submarines." Battleships are not mentioned anywhere in the act's language.

2. Miller, *War Plan Orange*, 186–93.

3. U.S. Naval War College, "Remarks of President Naval War College Preliminary to Solving Operations Problem III-1935-SR, 20 February 1935," by Edward Kalbfus, Folder 1914-C, Box 69, RG4, Publications, 1915–77, NHC, 1–7.

4. Moore, interview by Mason, 542–52.

5. Hattendorf, Simpson, and Wadleigh, *Sailors and Scholars*, 148.

6. Entry of 1 February 1935, naval diary of Harry L. Pence, Series 4, Harry Pence Papers, 1893–1976, MSS 0144, Mandeville Special Collections Library, Geisel Library, University of California, San Diego; and *The Mission and Organization of the Naval War College, 1936–1937*, Folder 108, Box 7, Collection 619, Papers of Dewitt C. Ramsey, 1914–49, Operational Archives Branch, NHHC, 5–7.

7. BROWN appears in OP VII 38 and 39, and PURPLE appears in TAC VI 34SR and OP III 37SR.

8. Vlahos, "War Gaming," 13.

9. U.S. Naval War College, "Introductory Remarks, Tactics Course," by Roscoe C. MacFall, Folder 1898, Box 67, RG4, Publications, 1915–77, NHC, 4. Alfred Thayer Mahan called Nelson's memorandum of 9 October 1805 "memorable not only for the sagacity and comprehensiveness of its general dispositions, but even more for the magnanimous confidence with which the details of execution were freely instrusted [*sic*] to those upon whom they had to fall." From Mahan, *Life of Nelson*, 349–50.

10. Captain Kerrick commanded USS *Arizona* (BB-41) from June 1932 until September 1933. Stillwell, *Battleship Arizona*, 333.

11. Captain Blakely commanded USS *Lexington* (CV-2) from May 1932 until June 1934. From "Blakely," *Dictionary of American Naval Fighting Ships*, accessed 14 April 2013, http://www.history.navy.mil/danfs/b7/blakely-iii .htm.

12. U.S. Naval War College, "Tactical Problem II-1935-SR; Notes Taken at Critique," Folder 1907-G, Box 68, RG4, Publications, 1915–77, NHC, 2.

13. Moore, interview by Mason, 521–22. Mervyn Bennion was posthumously awarded the Medal of Honor for his actions on 7 December 1941 as commanding officer of USS *West Virginia* (BB-48).

14. U.S. Naval War College, "Remarks of President Naval War College Preliminary to Solving Operations Problem III-1935-SR, 20 February 1935," 2, 7.

15. U.S. Naval War College, "Operations Problem III-1935-SR; BLUE Special Situation, Folder 1914-A, Box 69, RG4, Publications, 1915–77, NHC, 5; U.S. Naval War College, "Annex 1: CNO-CinC BLUE, Subj: Ops of BLUE Fleet in BLUE-ORANGE War," Folder 1914-A, Box 69, RG4, Publications, 1915–77, NHC, 8.

16. U.S. Naval War College, "Comments of the Landing Operations at Guam," enclosure E to Department of Intelligence, Naval War College, "Analysis of Operations Problem III-1935-SR and Tactical Problem IV," Folder 1914-H, Box 70, RG4, Publications, 1915–77, NHC, 1–2.

17. Department of Intelligence, U.S. Naval War College, "Analysis of Operations Problem III-1935-SR and Tactical Problem IV," Folder 1914-H, Box 70, RG4, Publications, 1915–77, NHC, 51.

18. U.S. Naval War College, "Tactical Problem IV-1935-SR," BLUE and ORANGE Battle Plans No. 1, Folder 1910-B, Box 69, RG4, Publications, 1915–77, NHC, 1, 2.

19. U.S. Naval War College, "Tactical Problem IV-1935-SR, Annex A and B, ORANGE Battle and Approach Dispositions," Folder 1910-B, Box 69, RG4 Publications, 1915–77, NHC. A similar formation was included in the statement of problem for TAC II-37. See U.S. Naval War College, "Tactical Problem II-1937-Senior; Schedule of Employment and Statement of Problem," Folder 2108, Box 79, RG4, Publications, 1915–77, NHC, 9.

20. Department of Intelligence, U.S. Naval War College, "Abstract of Trans-Pacific Problem, beginning 20 March 1935; OP. Prob III-Tac. Prob IV-1935-SR; Tactical Phase," Folder 1914-H, Box 70, RG4, Publications, 1915–77, NHC.

21. Department of Intelligence, U.S. Naval War College, "Analysis of Operations Problem III-1935-SR and Tactical Problem IV," Folder 1914-H, Box 70, RG4, Publications, 1915–77, NHC, 46, 48.

22. U.S. Naval War College, "Notes Taken at Critique," enclosure I to Department of Intelligence, Naval War College, "Analysis of Operations Problem III-1935-SR and Tactical Problem IV," Folder 1914-H, Box 70, RG4, Publications, 1915–77, NHC, 2.

23. U.S. Naval War College, *Register of Officers, 1884–1955*.

24. During the first months of the coming war, Draemel would serve as Admiral Nimitz's chief of staff until being relieved, ironically, by Rear Admiral Raymond Spruance.

25. Bernhard T. Bieri to George Dyer, "Recollections of R. K. Turner," 22 May 1976, Folder 45, Box 1, MS Coll. 37, Thomas Buell Papers, NHC.

26. "Tactical Problem II-1937-Senior; Schedule of Employment and Statement of Problem," Folder 2108, Box 79, RG4, Publications, 1915–77, NHC, 2.

27. Class statistics regarding achievement of flag rank were developed by John Hattendorf for Dr. Evelyn Cherpak, head of the NHC.

28. U.S. Naval War College, "Operations Problem VII (Strategic) 1938; BLUE Staff Solution," Folder 2166-I, Box 84, RG4, Publications, 1915–77, NHC, 41.

29. U.S. Naval War College, "Operations Problem VII," 3–7.

30. U.S. Naval War College, "Operations Problem VII (Strat.) 1938-SR-JR; Instructions for the Chart Maneuver and Critique," by J. W. Wilcox, Folder 2166-H, Box 84, RG4, Publications, 1915–77, NHC, 1.

31. U.S. Naval War College, "Operations Problem VII (Strategic) 1938; BLUE Staff Solution," Folder 2166-I, Box 84, RG4, Publications, 1915–77, NHC, 67–68.

32. U.S. Naval War College, "Operations Problem VII-1938; Blueprint Record of Maneuvers, Move 6," Folder 2166M, Box 84, RG4, Publications, 1915–77, NHC.

33. Bieri to Dyer, 22 May 1976.

34. U.S. Naval War College, "Operations Problem VII (Strategic); History of Maneuver," Folder 2166K, Box 84, RG4, Publications, 1915-77, NHC, 24-26.

35. U.S. Naval War College, "Operations Problem VIII (Tactical) 1938; BLUE Statement of the Problem," Folder 2167D, Box 84, RG4, Publications, 1915-77, NHC, 1-2.

36. The Two-Ocean Navy Bill called for a $1 billion expansion of the navy over ten years, which amounted to sixty-nine new ships. It restarted capital ship construction and authorized an increase in tonnage of forty thousand tons for aircraft carriers to join the three existing ones and two under construction to be completed by the spring of 1939.

37. Roosevelt, "Campaign Address at Madison Square Garden, New York City, October 28, 1940," in *Public Papers*, 502.

Conclusion

1. Buell, "Admiral Raymond A. Spruance and the Naval War College: Part I—Preparing for World War II," 32, and "Part II—From Student to Warrior," 33, 45.

2. Hart, *Strategy*, 321.

3. Five battleships were sunk at Pearl Harbor, but three (*Tennessee, Maryland* and *Pennsylvania*) were less damaged and able to sortie two weeks after the attack. They were joined within two months by *Colorado, New Mexico, Idaho*, and *Mississippi* and grouped into Battleship Division 1 under Vice Admiral William Pye. While the carrier task forces and their cruiser escorts absorbed the brunt of the naval war through 1942 and into 1943, the prewar battleships did not enter a combat zone until after each had completed a comprehensive overhaul.

4. Hornfischer, *Neptune's Inferno*, 22, 383.

5. Prange, Goldstein, and Dillon, *Miracle at Midway*, 59.

6. Hustvedt, interview, 186.

7. This figure does not include demonstration games, reenactment games, or quick-decision exercises.

8. U.S. Naval War College, "Operations Problem IV-1928-SR (Trans-Pacific problem)," Folder 1382-VI, Box 41, RG 4 Publications, 1915-77, NHC, 7 of Section (i) Conclusions and Lessons Learned, 3. The critique does not record the author's name, but it is most probably Captain John W. Greenslade.

Appendix A

1. Schifferle, *America's School for War*, 34.

2. Chester Nimitz and Roscoe MacFall, both graduates of the Naval Academy class of 1905, were promoted to commander while members of the War College class of 1923.

3. Willoughby, *Rum War at Sea*, 46-54.

4. Hattendorf, Simpson, and Wadleigh, *Sailors and Scholars*, 146-47.

Appendix C

1. Capital ship construction and service dates in this appendix are extracted from Breyer, *Battleships and Battle Cruisers*.

2. Office of Naval Intelligence, "Operations—Japanese Navy in the Indian and Pacific Oceans during War, 1914–1918," RG 45, Subject File, 1911–27, WA-5 Japan, Box 703, Folder 10, NND 913005, 98.

3. Office of Naval Intelligence, "Operations," 11, 22.

4. Saxon, "Anglo-Japanese Naval Cooperation," 62.

5. Ferrell, *American Diplomacy*, 518.

6. "Five Power Treaty on Naval Limitation, February 6 1922," World War II Resources, accessed 27 March 2011, www.ibiblio.org/pha/pre-war/1922/nav_lim.html.

7. Preston, *Illustrated History*, 11.

8. "Five Power Treaty on Naval Limitation, February 6 1922."

9. "Large Naval Ships Built in U.S. Shipyards: Battleships," Shipbuilding History, accessed September 15, 2015, http://www.shipbuildinghistory.com/history/navalships/battleships.htm; "Large Naval Ships Built in U.S. Ship-yards: Cruisers," Shipbuilding History, accessed September 15, 2015, http://www.shipbuildinghistory.com/history/navalships/cruisers.htm.

10. Evans and Peattie, *Kaigun*, 236.

11. Preston, *Cruisers*, 150.

12. Preston, *Cruisers*, 71–72.

13. Ferrell, *American Diplomacy*, 523.

Bibliography

Archival Collections and Unpublished Sources

Buell, Thomas J. Interview by Paul Stilwell, 2005. Transcript. Oral History Collection. U.S. Naval Institute, Annapolis MD.

Bieri, Bernhard H. Interview by John T. Mason Jr., 1970. Transcript. Oral History Collection. U.S. Naval Institute, Annapolis MD.

Chief of Naval Operations, Strategic Plans Division (1940–70) and Predecessor Organizations (1912–47), COLL/369. National Archives and Records Administration, College Park MD.

Deyo, Morton L. Papers. Naval Historical Foundation Collection. Manuscript Division, Library of Congress, Washington DC.

Dring, Walter, Jr. Interview by Anthony S. Nicolosi, 21 February 1975. Transcript. Oral History Collection. Naval War Gaming Project, U.S. Naval War College, Newport RI.

Gaudet Philip R. Interview by Francis J. McHugh, 7 September 1974. Transcript. Oral History Collection. Naval War Gaming Project, U.S. Naval War College, Newport RI.

Greenslade, John Wills. Papers. Naval Historical Foundation Collection. Manuscript Division, Library of Congress, Washington DC.

Hart, Thomas C. Papers, 1899–1960. Operational Archives Branch, Naval Historical Center, Washington Navy Yard DC.

Hustvedt, Olaf M. Interview, 1975. Transcript. Oral History Collection. U.S. Naval Institute, Annapolis MD.

Kimmel, Husband E. Papers. American Heritage Center, University of Wyoming, Laramie WY.

Luce, Stephen Bleeker. Papers. Naval Historical Foundation Collection. Manuscript Division, Library of Congress, Washington DC.

Martin, John K. Interview by Francis J. McHugh, 21 September 1974. Transcript. Oral History Collection. Naval War Gaming Project, U.S. Naval War College, Newport RI.

Moore, Charles Johnes. Interview by John T. Mason, 1967. Transcript. Naval History Project, Columbia Center for Oral History, Columbia University, New York NY.

Naval Historical Collection (NHC). Record Group 4, Publications, 1915–77; Record Group 29, Students, Staff and Faculty; Record Group 35, Naval War Gaming, 1916–2003. Naval War College, Newport RI.

Naval Records Collection of the Office of Naval Records and Library. Record Group 45.5.1. Office of Naval Intelligence.

Naval War College. "Outline History of the Naval War College, 1884 to 1937." Unpublished compilation. Naval War College Library, Newport RI.

Naval War College Photo Archives. Naval War College Museum, Newport RI.

Nimitz, Chester W. Papers. Operational Archives Branch, Naval Historical Center, Washington DC.

Pence, Harry. Papers. MSS 144. Mandeville Special Collections Library, University of California, San Diego.

Powell, Paulus P. Papers. Operational Archives Branch, Naval Historical Center, Washington DC.

Pratt, William V. Papers. Operational Archives Branch, Naval Historical Center, Washington DC.

Ramsey, Dewitt C. Papers. Operational Archives Branch, Naval Historical Center, Washington DC.

Records of the Joint Board, 1903–47. National Archives Microfilm Publication M1421, Record Group 225. National Archives and Records Administration, College Park MD.

Roosevelt, Theodore. Papers. Manuscript Division, Library of Congress, Washington DC.

Sims, William Sowden. Papers. Naval Historical Foundation Collection. Manuscript Division, Library of Congress, Washington DC.

Published Sources

"Admiral at the Front." *Time Magazine*, 24 November 1941.

Andrade, Ernest, Jr. "The Ship That Never Was: The Flying-Deck Cruiser." *Military Affairs* 32, no. 3 (December 1968): 132–40.

Barns, S. M., W. M. Kaufman, and H. T. Gannon. *The U.S. Naval War College: A Staff Study of Its Historical Background, Mission, and Educational Philosophy, Principles, and Concepts*. Newport RI: Naval War College Press, 1954.

Belknap, Reginald R. *The Yankee Mining Squadron; or, Laying the North Sea Mine Barrage*. Annapolis MD: United States Naval Institute, 1920.

Blair, Clay, Jr. *Silent Victory: The U.S. Submarine War against Japan*. Philadelphia PA: Lippincott, 1975.

Borneman, Walter R. *The Admirals: Nimitz, Halsey, Leahy, and King—The Five-Star Admirals Who Won the War at Sea*. New York: Little, Brown, 2012.

Borowski, Harry R. *Military Planning in the Twentieth Century*. Washington DC: Office of Air Force History, 1986.

Breyer, Siegfried. *Battleships and Battle Cruisers, 1905-1970*. Garden City NY: Doubleday, 1973.

Brewer, Garry D., and Martin Shubik. *The War Game: A Critique of Military Problem Solving*. Cambridge MA: Harvard University Press, 1979.

Buell, Thomas B. "Admiral Raymond A. Spruance and the Naval War College: Part I—Preparing for World War II." *Naval War College Review 23*, no.7 (March 1971): 31-51.

———. "Admiral Raymond A. Spruance and the Naval War College: Part II—From Student to Warrior." *Naval War College Review 23*, no. 8 (April 1971): 29-53.

———. *The Quiet Warrior: A Biography of Admiral Raymond A. Spruance*. New York: Little, Brown, 1974.

Bywater, Hector. *The Great Pacific War: A History of the American-Japanese Campaign of 1931-1933*. Bedford MA: Applewood Books, 2002.

Caffrey, Matthew. "Toward a History-Based Doctrine for Wargaming." *Air and Space Power Journal* (April 27, 2000): 33-56.

Cannon, James W. *War Gaming: A Sound Procedure for Testing Military Plans and Concepts*. Carlisle PA: Army War College, 1969.

Cherpak, Evelyn. *A Guide to Archives, Manuscripts, and Oral Histories in the Naval Historical Collection*. Newport RI: Naval War College, 1985.

Chisholm, Donald. *Waiting For Dead Men's Shoes: Origins and Development of the U.S. Navy's Officer Personnel System, 1793-1941*. Stanford CA: Stanford University Press, 2001.

Cline, Ray S. *Washington Command Post: The Operations Division*. Washington DC: Center of Military History, United States Army, 1990.

Cohen, Eliot A., and John Gooch. *Military Misfortunes: The Anatomy of Failure in War*. New York: Free Press, 1990.

Collins, John M. *Military Strategy: Principles, Practices, and Historical Perspectives*. Washington DC: Brassey's, 2001.

Department of the Navy, Bureau of Yards and Docks. *Building the Navy's Bases in WWII*. Vol. 1 of *History of the Bureau of Yards and Docks, and the Civil Engineering Corps, 1940-46*. Washington DC: Government Printing Office, 1947.

Dickenson, Frederick R. *War and National Reinvention: Japan in the Great War, 1914-1919*. Harvard East Asian Monographs, no. 177. Cambridge: Harvard University Press, 1999.

Downs, James F. *Naval Personnel Organization: A Cultural-Historical Approach*. Office of Naval Research report no. 0001 AD. Washington DC: Department of the Navy, 1982.

Doyle, Michael K. "The U.S. Navy and War Plan Orange, 1933-1940: Making Necessity a Virtue." *Naval War College Review 32*, no. 3 (May-June 1980): 49-63.

Dyer, George Carroll, VADM, USN (ret.). *The Amphibians Came to Conquer: The Story of Admiral Richmond Kelly Turner*. Washington DC: Government Printing Office, 1969.

Ellis, Earl H. *Advanced Base Operations in Micronesia, FMFRP 12-46.* Washington DC: Department of the Navy, Headquarters United States Marine Corps, 1992.

Eure, Leroy T. *Development of War Gaming in the Naval War College, 1886–1958.* Newport RI: Naval War College Press, 1959.

Evans, David C., and Mark R. Peattie. *Kaigun—Strategy, Tactics, and Technology in the Imperial Japanese Navy, 1887-1941.* Annapolis MD: Naval Institute Press, 1997.

Felker, Craig. C. *Testing American Sea Power: U.S. Navy Strategic Exercises, 1923-1940.* College Station: Texas A&M University, 2007.

Ferrell, Robert H. *American Diplomacy: A History.* New York: W. W. Norton, 1975.

Frank, Richard B. "Picking Winners?" *Naval History* 25, no. 3 (June 2011):24-32.

Friedman, Hal M. *Digesting History: The Naval War College, Lessons of World War II, and the Future of Naval Warfare.* Newport RI: Naval War College Press, 2010.

Furer, Julius Augustus. *Administration of the Navy Department in World War II.* Washington DC: Government Printing Office, 1959.

Gardiner, Robert, and Randal Gray. *Conway's All the World's Fighting Ships, 1906-1921.* London: Conway Maritime Press, 1986.

Gleeves, Albert. *Life and Letters of Rear Admiral Stephen B. Luce.* New York: Putnam, 1925.

Gole, Henry G. *The Road to Rainbow: Army Planning for Global War, 1934–1940.* Annapolis MD: Naval Institute Press, 2003.

Goodspeed, Hill. "The Interwar Transformation." *Naval History* 25, no. 3 (June 2011): 12-15.

Greenberg, Abe. "An Outline of Wargaming." *Naval War College Review* 34, no. 5 (September–October 1981): 93-97.

Greenfield, Kent Roberts, ed. *Command Decisions.* Washington DC: Center for Military History, 1960.

Groome, Francis Hindes. *Kriegspiel: The War Game.* New York: Ward, Lock and Bowden, 1896.

Halsey, Fleet Admiral William F., Jr., USN, and LCDR J. Bryan III, USNR. *Admiral Halsey's Story.* New York: McGraw Hill, 1947.

Hattendorf, John B., and Lynn C. Hattendorf, eds. *A Bibliography of the Works of Alfred Thayer Mahan.* Newport RI: Naval War College Press, 1986.

Hattendorf, John B., B. Mitchell Simpson III, and John R. Wadleigh. *Sailors and Scholars: The Centennial History of the U.S. Naval War College.* Newport RI: Naval War College Press, 1984.

Hone, Thomas C., and Trent Hone. *Battle Line: The United States Navy, 1919–1939.* Annapolis MD: Naval Institute Press, 2006.

Hornfischer, James D. *Neptune's Inferno: The U.S. Navy at Guadalcanal.* New York: Bantam Books, 2011.

Hoyt, Edwin P. *How They Won the War in the Pacific: Nimitz and His Admirals.* Guilford CT: Lyons Press, 2002.

Hughes, Wayne P., Jr. *Fleet Tactics: Theory and Practice.* Annapolis MD: Naval Institute Press, 1986.

——, ed. *Military Modeling.* Alexandria VA: Military Operations Research Society, 1984.

Jane, Fred T. "The Naval War Game." U.S. Naval Institute *Proceedings* 29, no. 3 (September 1903): 595–660.

Karsten, Peter. *The Naval Aristocracy: The Golden Age of Annapolis and the Emergence of American Navalism.* New York: Free Press, 1972.

Kennedy, Gerald John. *United States Naval War College, 1919–1941: An Institutional Response to Naval Preparedness.* PhD diss., University of Minnesota, June 1975.

Kimmel, Husband A. *Admiral Kimmel's Story.* Chicago: H. Regnery, 1955.

King, Ernest J., and Walter Muir Whitehill. *Fleet Admiral King.* New York: W. W. Norton, 1952.

Kirschbaum, Joseph W. *The 1916 Naval Expansion Act: Planning for a Navy Second to None.* PhD diss., George Mason University, 31 August 2008.

Kuehn, John T. *Agents of Innovation: The General Board and the Design of the Fleet That Defeated the Japanese Navy.* Annapolis MD: Naval Institute Press, 2008.

Laning, Harris. *An Admiral's Yarn.* Newport RI: Naval War College Press, 1999.

Lea, Homer. *The Day of the Saxon.* New York: Harper & Bros., 1912.

LaFeber, Walter. *The Clash: A History of U.S.-Japan Relations.* New York: W. W. Norton, 1997.

Leutze, James. *A Different Kind of Victory: A Biography of Admiral Thomas C. Hart.* Annapolis MD: Naval Institute Press, 1981.

Liddell Hart, Basil H. *Strategy.* 2nd ed. London: Faber, 1967.

Little, W. McCarty. "The Strategic Naval War Game or Chart Maneuver." U.S. Naval Institute *Proceedings* 38, no. 4 (December 1912): 1213–34.

Lowenthal, Mark M. *Leadership and Indecision: American War Planning and Policy Process, 1937–42.* 2 vols. New York: Garland, 1988.

MacDonald, Scot. *Evolution of Aircraft Carriers.* Washington DC: Office of the Chief of Naval Operations, 1962.

Mahan, Alfred Thayer. *The Influence of Sea Power of History, 1660–1783.* Boston: Little, Brown, 1890.

——. *The Life of Nelson: The Embodiment of the Sea Power of Great Britain.* Vol. 2. London: Sampson Low, Marston, 1897.

——. "The Necessity and Object of a Naval War College: Address of Captain A. T. Mahan, U.S. Navy, at the Opening of the Fourth Annual Session of the College, August 6, 1888." U.S. Naval Institute *Proceedings* 14, no. 4 (1888):621–39.

Mandel, Paul. "The World's Most Complicated Game." *Sports Illustrated,* September 23, 1963.

Matloff, Maurice, and Edwin M. Snell. *Strategic Planning for Coalition War-fare, 1941-1942*. Washington DC: Center of Military History, 1990.

McBride, William M. *Technological Change and the United States Navy, 1865-1945*. Baltimore: Johns Hopkins University Press, 2000.

McCue, Brian. *Wotan's Workshop: Military Experiments before WWII*. Alexandria VA: CNA, 2005.

McHenry, Robert, ed. *Webster's American Military Biographies*. Springfield MA: G. & C. Merriam, 1978.

McHugh, Francis J. "Eighty Years of War Gaming." *Naval War College Review* 21, no. 7 (March 1969): 88-90.

———. *Fundamentals of War Gaming*. Newport RI: Naval War College Press, 1966.

Miller, Edward S. *War Plan ORANGE: The U.S. Strategy to Defeat Japan, 1897-1945*. Annapolis MD: Naval Institute Press, 1991.

Morison, Samuel Eliot. *Rising Sun in the Pacific*. Vol. 3 of *History of the United States Naval Operations in World War II*. New York: Little, Brown, 1948.

Murray, Williamson R., and Allan R. Millett, eds. *Military Innovation in the Interwar Period*. New York: Cambridge University Press, 1996.

Nofi, Albert A. *To Train the Fleet for War: The U.S. Navy Fleet Problems, 1923-1940*. Newport RI: Naval War College Press, 2010.

O'Neill, Robert. "Churchill, Japan, and British Security in the Pacific, 1904-1942." In *Churchill: A Major New Assessment of His Life in Peace and War*, edited by Robert Blake and William Roger Lewis, 275-90. Oxford: Clarendon Press, 1993.

Pennington, Leon Alfred, Jr., Romeyn B. Hough, and H. W. Case. *The Psychology of Military Leadership*. Whitefish MT: Kessinger, 2007.

Perla, Peter P. *The Art of Wargaming: A Guide for Professionals and Hobbyists*. Annapolis MD: U.S. Naval Institute, 1990.

Perla, Peter, and Ed McGrady. "Why Wargaming Works." *Naval War College Review* 64, no. 3 (Summer 2011): 111-30.

Pois, Robert, and Phillip Langer. *Command Failure in War: Psychology and Leadership*. Bloomington: Indiana University Press, 2004.

Potter, E. B. *Nimitz*. Annapolis MD: U.S. Naval Institute Press, 1976.

Prados, John. *Pentagon Games: Wargames and the American Military*. New York: Harper and Row, 1987.

Prange, Gordon W. *December 7th, 1941*. New York: Warner Books, 1988.

Prange, Gordon W., Donald M. Goldstein, and Katherine V. Dillon. *Miracle at Midway*. New York: Penguin Books, 1983.

Pratt, William Veazie. "The Naval War College: Outline of the Past and Description of the Present." U.S. Naval Institute *Proceedings* 53, no. 9 (September 1927): 937-47.

Preston, Anthony. *Cruisers—An Illustrated History, 1880-1980*. London: Bison Books, 1980.

———. *An Illustrated History of the Navies of World War II*. London: Bison Books, 1976.

Richardson, James O. *On the Treadmill to Pearl Harbor: The Memoirs of James O. Richardson USN (Retired)*. Washington DC: Naval History Division, Department of the Navy, 1973.

Roosevelt, Franklin, D. *The Public Papers and Addresses of Franklin D. Roosevelt, 1940*. Vol. 9, *War and Aid to Democracies*. New York: Macmillan, 1940.

Rosen, Stephen Peter. "New Ways of War: Understanding Military Innovation." *International Security* 13, no. 1 (Summer 1988): 134–69.

———. *Winning the Next War: Innovation and the Modern Military*. Ithaca NY: Cornell University Press, 1991.

Saxon, Timothy D. "Anglo-Japanese Naval Cooperation, 1914–1918." *Naval War College Review* 53, no. 1 (Winter 2000): 62–93.

Schifferle, Peter J. *America's School for War: Fort Leavenworth, Officer Education, and Victory in World War II*. Lawrence: University Press of Kansas, 2010.

Smith, Douglas V. "Preparing for War: Naval Education between the World Wars." *International Journal of Naval History*, no. 1 (April 2002).

Spector, Ronald. *Eagle against the Sun: The American War with Japan*. New York: Free Press, 1985.

———. *Professors of War: The Naval War College and the Development of the Naval Profession*. Newport RI: Naval War College Press, 1977.

Still, William N., Jr., ed. *The Queenstown Patrol, 1917: The Diary of Joseph Knefler Taussig, U.S. Navy*. Newport RI: Naval War College Press, 1996.

Stillwell, Paul. *Battleship Arizona: An Illustrated History*. Annapolis MD: Naval Institute Press, 1991.

Stoler, Mark A. *Allies and Adversaries: The Joint Chiefs of Staff, the Grand Alliance, and U.S. Strategy in World War II*. Chapel Hill: University of North Carolina Press, 2000.

Sturton, Ian, ed. *All the World's Battleships, 1906 to the Present*. London: Brassey's, 1996.

Todd, Forde A. *A Study and Discussion of the Estimate of the Situation*. Newport RI: U.S. Naval War College Department of Operations, 1933.

Toft, Monica, and Talbot Imlay, eds. *The Fog of Peace and War Planning: Military and Strategic Planning under Uncertainty*. New York: Routledge, 2006.

Totten, C. A. L. "Strategos, the American War Game." *Journal of the Military Service Institution* 1 (1880): 185–202.

U.S. Naval War College. *The Estimate of the Situation: Plans and Orders*. Newport RI: U.S. Naval War College, 1929.

———. *Register of Officers, 1884–1977*. Newport RI: U.S. Naval War College, 1977.

———. *Sound Military Decision, Including the Estimate of the Situation and the Formulation of Directives*. Newport RI: U.S. Naval War College, 1936.

U.S. War Department. *The Report of the Secretary of War to the President, 1927*. Washington DC: Government Printing Office, 1927.

Vlahos, Michael. *The Blue Sword: The Naval War College and the American Mission, 1919–1941*. Newport RI: Naval War College Press, 1980.

———. "Wargaming: An Enforcer of Strategic Realism, 1919–1942." *Naval War College Review* 39, no. 2 (March–April 1986): 7–22.

Von Reisswitz, B. *Anleitung zur Darstellung militairischer Manöver mit dem Apparat des Kriegs-Spieles* [Instructions for the representation of military maneuvers with the apparatus of the war-game]. Berlin: Trowitzch, 1824.

Watson, Mark Skinner. *Chief of Staff: Prewar Plans and Preparations.* Washington DC: Center for Military History, 1990.

Weigley, Russell F. *The American Way of War: A History of United States Military Strategy and Policy.* New York: Macmillan, 1973.

Weir, Gary E. *Building American Submarines, 1914–1940.* Washington DC: Department of the Navy, Naval Historical Center, 1991.

Wheeler, Gerald E. *Admiral William Veazie Pratt, U.S. Navy: A Sailor's Life.* Washington DC: Naval History Division, Department of the Navy, 1974.

Wildenberg, Thomas. *All the Factors of Victory: Admiral Joseph Mason Reeves and the Origins of Carrier Air Power.* Dulles VA: Brassey's, 2003.

———. *Gray Steel and Black Oil: Fast Tankers and Replenishment at Sea in the U.S. Navy, 1912–1992.* Annapolis MD: Naval Institute Press, 1996.

Willmot, H. P. *Empires in the Balance: Japanese and Allied Pacific Strategies to April 1942.* Annapolis MD: Naval Institute Press, 1982.

Willoughby, Malcolm F. *Rum War at Sea.* Washington DC: Government Printing Office, 1964.

Withers, Thomas. "The Preparation of the Submarines Pacific for War." U.S. Naval Institute *Proceedings* 76, no. 4 (April 1950): 387–93.

Wylie, J. C. *Military Strategy: A General Theory of Power Control.* Annapolis MD: Naval Institute Press, 1989.

Index